TO THE RED PLANET

TO THE RED PLANET

Eric Burgess

Columbia University Press New York 1978

Eric Burgess is a free-lance writer
who has written extensively on the
upper atmosphere, rockets, and
space flight.

Library of Congress Cataloging in Publication Data

Burgess, Eric.
    To the red planet.

    Includes index.
    1. Mars (Planet)—Exploration.   2. Viking Mars
Program.   I. Title.
QB641.B82   919.9'23'04   78-6911
ISBN 0-231-04392-9

Columbia University Press
New York and Guildford, Surrey
Copyright © 1978 Columbia University Press
All rights reserved
Printed in the United States of America

ACKNOWLEDGMENTS
Unless otherwise indicated in the captions, all figures and pictures are from the National Aeronautics and Space Administration: Particular thanks are due to the Public Information Offices of NASA's Langley and Ames Research Centers, and the Jet Propulsion Laboratory generally and to Gary Price, Stanley Miller, and Frank Bristow at the respective facilities. Thanks are also gratefully given to Don Bane for his help when at TRW Systems during the nominal Viking mission and at the Jet Propulsion Laboratory during the extended mission.

Contents

Mars in the Human Mind

As thinking beings, men and women seem to have always accepted the idea that other intelligences, other consciousnesses, might be "out there" in the universe.

Early in recorded history mankind believed in pantheons of gods and in hordes of mythical creatures able to travel beyond the confines of the earth as it was then known. Heavenly bodies were associated with intelligences. They were regarded as beings separate from but interested in the men and women of the earth, or as the abodes of gods.

When astronomers applied telescopes and, later, other instruments to learn about the heavenly bodies, when they discovered more about the physical nature of the moon and the planets in the solar system, human thought changed extensively. However, people still nourished the idea that there existed other living things, extraterrestrial life forms.

In the early nineteenth century the readiness of people to believe in life on other worlds was demonstrated by the great Herschel hoax. It began as a satire by a writer at the New York *Sun,* designed to catch the popular imagination and increase the *Sun*'s circulation. The series certainly did get attention, but people took it as literal truth. At that time it was fairly

well known that Sir John Herschel, son of the man who discovered Uranus (Sir William Herschel), had begun observing southern skies with a large new telescope located at Capetown, South Africa. The *Sun*'s reporter concocted a story about an article that he claimed Herschel had written for an Edinburgh science journal. A series of "exclusive" reports began with details of the new telescope. Subsequent installments of the story described a light intensifier that allowed tremendous magnifications. With these wonderful instruments, it was reported, Herschel saw buildings on the moon, and animals, men, and women. Circulation increased enormously, and the *Sun* kept the series going until the hoax was finally exposed.

Despite people's undoubted interest in possibilities of life beyond the earth, Mars was ignored as a possible inhabited planet until the middle of the nineteenth century, when astronomers showed that the moon could not be inhabited. Then attention gradually turned to Mars as a possible abode for other intelligent beings. Speculation could run largely unchecked by fact because, although telescopes revealed intricate details of the moon, they produced only hazy and indistinct views of Mars. The moon, with a diameter of 2,160 miles (3,476 kilometers), shows a disc 100 times larger than that of distant

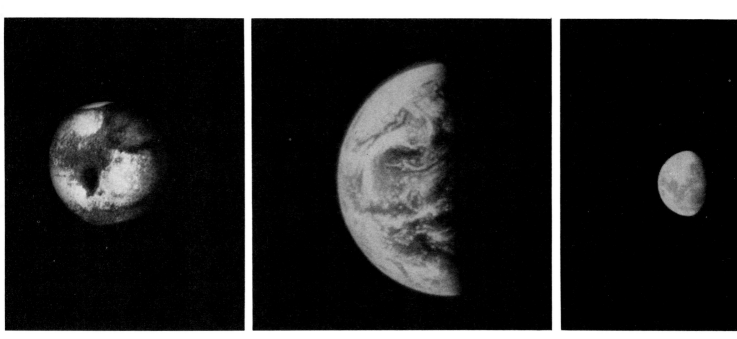

1.1 Seen in a telescope from earth, Mars exhibits an extremely small disc compared with the moon, as shown in *A*, a comparison photograph of part of the moon and the disc of Mars photographed at a close opposition. Part *B* shows Mars, earth, and moon in their true relative sizes.

1.2 A globe of Mars, showing Syrtis Major and Hellas, produced by Percival Lowell from his observations in 1896.

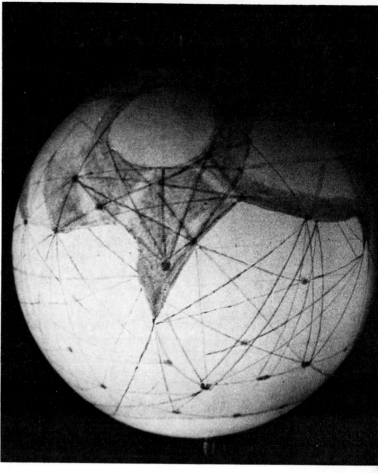

Mars (which has a diameter twice as great), even at Mars' closest approach to the earth (figure 1.1). At other times Mars is much smaller. In spite of Mars' small visual size, keen-eyed observers saw enough details through telescopes to allow them to draw maps of the planet. Such maps were often quite dissimilar, however, because of the personal styles used. Generally the maps agreed on the location of large areas of reddish-ocher color and darker areas of blues and greens. These markings seemed to be fairly consistent in general shape and position over the years, although subtle changes were observed and recorded. The planet also has light-colored polar caps which astronomers assumed to be of snow and ice. Many observers also assumed that the light-colored areas were deserts and the darker areas were vegetation. It was concluded that conditions for life on Mars did not differ too greatly from the earth. In fact, some early astronomers thought, mistakenly, that the dark areas were seas. Later it became commonly accepted that Mars was a drier and colder place than the earth, and people speculated that it might be a more advanced planet, with a higher degree of civilization.

After observations in 1877 by the Italian astronomer Giovanni Schiaparelli, the late 1800s became an era of belief in Martian "canals." (Shiaparelli referred to the linear markings as *canali*, which translates as "channels," not "canals.") Astronomers drew long straight lines crisscrossing the planet. This era culminated in Percival Lowell's assertion that the canals were the works of intelligent beings trying to distribute dwindling supplies of water over a dying and desiccated planet. Lowell's maps, made at an observatory he established in 1894 expressly to study Mars through the clear skies of Flagstaff, Arizona, showed geometrical patterns of lines across the planet. Some of these lines, he claimed, doubled at certain seasons (figure 1.2). Dark spots, at the intersection of the lines, he called oases; and he was convinced that the canals represented major engineering works of in-

telligent beings on the Red Planet. His view was strongly opposed by many other astronomers.

The possible existence of Martians intrigued many people, however. In 1892 Camille Flammarion published *La Planète Mars et ses conditions d'habitabilité*, in which he clearly regarded Mars as the abode of intelligent life. In the early 1900s, when Percival Lowell was certain that Mars possessed an advanced civilization, the acute observer Eugenios Antoniadi claimed that no fine lines were visible on Mars, nothing on which to base Flammarion's and Lowell's speculations about life on Mars. What Schiaparelli had drawn as geometrical lines, Antoniadi showed as small spots. The Martian canals, it seemed, were created by the human eye, joining details that were almost at the limits of visibility.

Photography did not help much until comparatively recently, when linear markings on Mars were obtained on earth-based pictures of the planet. Some

Mars in the Human Mind

linear markings are undoubtedly visible on the best modern photographs, but they do not correspond to geological features recorded on photographs of the Martian surface taken from spacecraft. One thing is certain—the "canals" are not artificial waterways built by intelligent Martians to carry water from the melting polar caps.

In fact, Lowell's speculation was disproved soon after it was first aired. The rapid disappearance of the seasonal polar caps showed that if they were water, they must be extremely thin and could not contain enough water to make such a hypothetical canal system worthwhile. Moreover, because of the low atmospheric pressure on Mars, it was most unlikely that free water could remain on the surface. Until quite recently, spectroscopic examination from earth did not reveal any water vapor in the Martian atmosphere. Mars is, indeed, a very dry world. Most estimates of the temperature of Mars concluded that it is also a cold world compared with the earth, always at least several tens of degrees below freezing. But the fascination of narrow straight markings on Mars lasted for a long while.

Several nineteenth-century events encouraged speculation about extraterrestrial life. Friedrich Wohler demonstrated in 1828 that living things contain the same basic materials as nonliving objects. About the middle of the century the spectroscope, invented by Gustav Kirchoff and Robert Bunsen, was used to show that other celestial bodies contain the same chemical elements as the earth. Biologists stated that all life on earth is interrelated.

The hypothesis of Immanuel Kant and Pierre Laplace for the formation of the solar system implied that Mars was an older planet than the earth. So, in view of Charles Darwin's theories of evolution, people naturally speculated that a civilization on Mars could have developed and evolved to a much more advanced form than existed on the younger earth.

As a result, Mars as an abode of life began to appear in fiction more frequently. In the late 1890s, Kurd Lasswitz in Germany wrote a Martian novel entitled *Between Two Planets,* and a few years later H. G. Wells wrote *The War of the Worlds,* describing an invasion of earth by Martians who possess a very advanced technology but are unable to resist terrestrial microbes. As recently as 1938, Orson Welles caused many people to panic as a result of a CBS radio broadcast adaptation of the H. G. Wells story, which he arranged in the form of a newscast on Halloween.

My first view of Mars was a red star rising in the eastern sky below Denebola, the second brightest star of the constellation Leo. The time was 2:00 A.M. on December 27, 1934. It was a cold night in England, and I had stayed up late for a family reunion at the home of a relative. As a high-school student I had been interested in astronomy for about a year. I had already looked at the moon through a small home-made telescope. I was thrilled to see the reddish star in the crisp winter night, and hurried to get my telescope. But while it had revealed craters and plains on the moon, the telescope showed Mars only as a disappointingly fuzzy, indistinct blob of light.

In subsequent months I continued to watch Mars as it moved eastward into the constellation Virgo. On March 13, the Red Planet began its great retrograde loop that lasted until May. This reversal to an apparent westward movement through the stars occurs around each opposition (the time when the earth is directly between Mars and the sun). It is caused by the earth overtaking the more slow-moving Mars, whose orbit is farther from the sun. Mars attained its brightest magnitude about the time of opposition on April 6, 1935, when it was 57 million miles (91 million km) from earth and exactly opposite the sun as seen from earth.

During this period Father J. P. Rowland of Stoneyhurst College, a well-known British astronomer and seismologist, introduced me to the Manchester Astronomical Society, which was associated with the College of Technology of that city. On May 16, 1935, E. Denton Sherlock, a senior member of the Society, invited me to his home to discuss a telescope mirror that I was making. The night was clear, so he invited me to view Mars through his 12-inch (30.5-cm) reflector. For the first time I saw the bright polar cap of Mars and some indistinct dark markings on the reddish globe.

Through this telescope, the surface of Mars exhibited bright reddish-ocher regions delicately tinged with shades of pink. It is these relatively large areas on the planet's surface that produce the characteristic red color of Mars, so that it appears as a red star in the skies of earth, and earns its name of the "Red Planet."

I could just make out some of the darker areas of gray, green, and blue that early astronomers thought were seas. They had quickly realized that these could not be large bodies of water, however, because they did not reflect sunlight like water. The changes in these dark areas, in what appeared to be seasonal cycles, had led some astronomers to speculate that they were covered with vegetation.

By the time I first saw Mars in a telescope, I had borrowed Lowell's books from the library of the Society and was steeped in Mars lore. I had also located a copy of Alfred Russel Wallace's book, *Is Mars Habitable?* in a used book store. Wallace attacked Lowell's speculations. I was impressed because Wallace, along with Charles Darwin, had discovered the theory of natural selection. Gradually I became emotionally involved in the life-on-Mars controversy as I sought more information about the two points of view. I was somewhat surprised that few of the amateur and professional astronomers I met at the Society's meetings showed an interest in Mars; there were very few planetary astronomers in the 1930s

because the realm of the nebulae dominated astronomical thought. To stir up interest I wrote an article entitled "Mars, Possibilities of Life" that was published in a local magazine called the *Manchester Forum*, in April 1936.

In the 40 years since then, in common with others of my generation, I have witnessed a complete change in attitudes about Mars. It is an experience shared with many people who read Edgar Rice Burroughs' Martian Series, H. G. Wells' *War of the Worlds*, and a plethora of wildly speculative science-fiction stories whose descriptions of Mars ranged from a densely populated Eden to a veritable desert.

Mars lore played a major role in stimulating me to help form an interplanetary society in England in the mid-1930s, although the initial aim of the society was to try to get to the moon. Getting men to Mars seemed an extremely remote possibility even to the incurable rocket optimists with whom I mingled. At that time, rocket engines the size of eggs could only be made to work for a short while before exploding.

After writing many technical articles during World War II on the engineering aspects of rockets, thermodynamics of reaction propulsion engines, and the like, and detailed technical reviews of rocket development in Germany, I again published an article on Mars, some 15 years after my first. It stemmed from a detailed technical article on the establishment and uses of unmanned artificial satellites.* It was, indeed, the first detailed paper to show the feasibility of earth satellites for scientific and communication purposes, based on technology available from World War II. It described the best way to attain a geosynchronous orbit, in which the satellite revolves around the earth in the same period of 24 hours that earth takes to rotate on its axis. Also, it discussed the

*"The Establishment and Uses of Artificial Satellites," *Aeronautics*, vol. 21 (September 1949), pp. 70–82.

stability of such an orbit and how solar energy might be tapped and used in space to relay television programs. I began thinking of what else might be done with unmanned rockets based on technology expected during the next decades.

In those days the interplanetary societies were still thinking actively about putting earth satellites in orbit as the first step in getting men to the moon, so I decided to look beyond the moon and check if there were possibilities of exploring Mars. Manned flights to Mars were out of the question with foreseeable technology, not because of the energy requirements but rather because the distance to Mars and the motions of the planets on their orbits would require a crew to be away from the earth and supported in space or on Mars for about 3 years. In the early 1950s there seemed little likelihood of perfecting life-support systems for such a long mission in space.

But what about an unmanned spacecraft for a one-way journey?

After several months of searching the literature and calculating orbits on the slow mechanical calculators of that time, I found that it was feasible, using the largest liquid-propellant rocket engines then available and the type of multistage rocket structures then visualized, together with solid-propellant modules for the initial thrust, to send an unmanned probe to Mars. The probe would orbit the planet and send back, by facsimile techniques like those used to transmit news photos, reasonably detailed pictures of Mars that might answer some of the questions about the Red Planet that could never be answered from earth.

My findings were published in *Aeronautics*,* and as a result the editor of the *Manchester Evening News*

* "The Martian Probe," *Aeronautics,* vol. 27 (November 1952), pp. 26–33.

asked for a popular account to use as a feature article. It soon appeared as a full-page story under the headline: "With a Robot Space Cruiser We Could Look In at Mars." Requests from engineering and nonprofessional groups then kept me happily talking about Mars probes for months. With C. Arthur Cross, I also gave a paper to the British Interplanetary Society on the Mars probe, which was published in the Society's *Journal* in March 1953.

About this time too, my friend Arthur C. Clarke sent me a copy of his new book *The Sands of Mars*. Clarke assumed in his story that Mars had been colonized, and humans were developing the planet. I remember well the passage where he described the arrival of the hero at Mars and how the view changed from a planet floating in space to a landscape and finally to a desert with hardy plants of mottled green; I wondered at the time if my Mars-probe idea could also provide such a change in human viewpoint about the Red Planet.

The idea of landing anything on the surface of Mars seemed only a dream, since it appeared unlikely that any government would ever finance such an expedition. We could visualize sending an unmanned spacecraft into Martian orbit, but a landing would be a great step farther in technology and in cost.

Yet in another quarter of a century the dream of landing successfully on Mars was realized when the first Viking spacecraft touched down on the reddish-ocher desert of Chryse on July 20, 1976. (Soviet spacecraft had landed earlier, but had failed to operate on the surface of Mars.) This NASA expedition to Mars involved an unmanned machine of enormous complexity far beyond the wildest dreams of the visionaries of the 1950s, who had a hard time trying to convince governments that earth satellites were worthwhile. The dream was not made possible by the development of big rockets as we had visual-

Mars in the Human Mind

ized them in the early days. In fact, the launch vehicles for the first Mars probes and the Mars lander were small compared with those used in landing men on the moon. But the dream was made a reality by the developments that were relatively new to the interplanetary movement in the 1950s as I recall it. The breakthroughs came from microminiaturization of digital computers and of biological experiments, rather than from the brute force of propulsion systems.

The Viking expedition to explore Mars depended on electronics rather than rocketry: electronics to perform orbital calculations and to navigate the spacecraft to a safe landing on a distant world through an atmosphere containing unknown gases; electronics to control the spacecraft, to orient it, to direct the operation of its many grosser systems; electronics to convert the energy of sunlight into electricity for an orbiter and nuclear decay into electricity for a lander. It was sophisticated electronics that produced fantastic, clear pictures from orbit and from the surface, controlled a complex biological laboratory reduced to one cubic foot in size, sensed the Martian winds, sniffed the Martian air, recorded the shaking of the Martian crust. And electronics in the spacecraft and on earth passed immense quantities of information across hundreds of millions of miles of space to tell us here on earth what it is like on the surface of Mars. Finally, electronics also let the public participate in the exploration of Mars through the TV screens in their own homes.

Such electronics provided an extension of our senses whereby we could, in effect, step out upon and look around the world of Mars. This sensory extension offered the stimulating opportunity to answer the question of whether or not Mars was, indeed, the abode of life. We could look around on the surface for plants, for evidence of living creatures, for ruins of ancient civilizations. And through the advanced technology of microminiaturization we could even carry a complex biological laboratory to Mars to inspect the soil of the Red Planet for evidence of microorganisms, dead or alive.

One of the Martian features that encouraged speculation about life was the color of the dark areas. Percival Lowell claimed that these areas showed a progressive change of color with the Martian seasons; over a period of months, green changed gradually to brown in a sequence that was repeated each year (figure 1.3). Some of the areas, such as the Solis

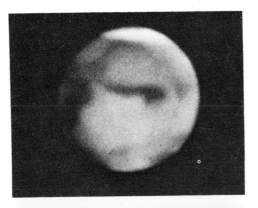

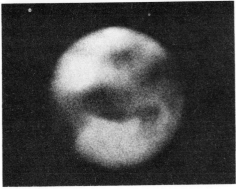

1.3 Astronomers claimed that a "wave of darkening" swept across Mars as the polar caps shrank each summer. These two photographs illustrate this darkening of prominent features between spring (*top*) and summer (*bottom*) in the southern hemisphere of Mars. These pictures are oriented with south at the top as seen in an astronomical telescope.

(PHOTO, LOWELL OBSERVATORY)

Lacus (Eye of Mars), changed considerably over the years, while others followed a repeating pattern of changes from one Martian year to the next. Other observers claimed that the dark areas were only grayish, and their green color was merely an effect of contrast with the ocher regions.

One of the most telling arguments against the dark areas being vegetation was that they did not reflect sunlight in the way that terrestrial green plants did. An alternative theory was that the color changes were caused by hygroscopic (water-attracting) salts that absorbed water vapor as the polar caps melted and released moisture into the Martian atmosphere.

An atmosphere on Mars was recognized very soon after the first telescopic observations of the planet. Vague mists and obscurations of surface markings were interpreted as due to clouds, which implied the presence of an atmosphere. As long ago as 1784, Sir William Herschel accepted the idea of clouds and vapors on Mars. But almost a century passed before attempts were made to determine the nature of the atmosphere using the newly invented spectroscope, which dispersed the light from Mars into a spectrum on which dark lines were produced by absorbing gases in the atmospheres of Mars and the earth. These attempts were singularly unsuccessful. Despite expeditions to mountain tops to make observations at heights that would reduce the effects of earth's atmosphere, astronomers were unable to detect water vapor or oxygen on Mars. As a consequence the Martian atmosphere was believed to consist of carbon dioxide and nitrogen. The lack of measurable amounts of water vapor also caused speculation that the polar caps could not be water ice but were frozen carbon dioxide.

Estimates of Martian atmospheric pressure placed it at 10 percent of the earth's average surface pressure of 1,013 millibars. By the mid-1950s the best es-timates were that the atmosphere of Mars was about 95 percent nitrogen and the rest carbon dioxide, with traces of argon, water vapor, and oxygen. Only carbon dioxide had been positively identified; nitrogen cannot be observed spectroscopically from earth because it does not produce any features in the observable region of the spectrum, and its abundance was estimated by analogy with the earth's atmosphere.

Since Mars is farther from the sun than is the earth, some astronomers regarded the planet as a cold world. Others argued that the melting of the polar caps in summer proved that Mars became quite warm. But heat radiation measurements made in the 1920s revealed a noon equatorial temperature on Mars of about freezing, and the temperature of the polar caps as −94 degrees Fahrenheit (−70° Celsius), rising to −54 degrees Fahrenheit (−48°C) in midsummer at the pole. Subsequently much higher temperatures were measured, up to 80 degrees Fahrenheit (27°C) at the center of the planetary disc viewed from earth.

As the years passed and methods of observation and measurement improved, both the temperature and the air pressure on Mars plummeted. By the 1970s the atmospheric pressure on Mars had been established as less than 1 percent of earth's, and daytime maximum temperatures were believed to be only about 32 degrees Fahrenheit (0°C) in midsummer, plunging to −135 degrees Fahrenheit (−93°C) at night.

While possibilities of life on Mars received a tremendous setback because of the physical conditions there, they bounced back into favor because field and laboratory experiments on earth showed the remarkable flexibility of terrestrial life. Biologists found that life on earth had adapted naturally to a large range of temperatures, pressures, and environments. They found that some forms of terrestrial life could also adapt to simulated Martian conditions. Surely, thought many scientists, if life evolved on earth and on Mars (when ages ago it was more like earth), such life could have adapted to the changing conditions on Mars. Life might still be present on the desiccated and almost airless Red Planet of today.

The question of life on Mars had had many ups and downs since 1659, when Christian Huygens first observed the triangular dark marking named Syrtis Major; the answer was still "maybe" when two Viking spacecraft were launched.

Biologists developing the instruments to check for life on Mars did not hold much hope of finding life there. In January 1973 I talked with Dr. Norman Horowitz, a geneticist at the California Institute of Technology in Pasadena, about possibilities of extraterrestrial life. He thought that

> Mars is marginal for life, [but] . . . far more interesting today than it was before Mariner 9 [a Mariner spacecraft that orbited Mars in 1971]; it is a much more interesting planet, biologically, geologically, and every other way. I think the findings do encourage, to some extent, our expectations of finding life on Mars, but I still find Mars to be a very hostile place compared to any place on earth. I think there's a chance for life on Mars, but I don't think it's a strong chance.

I also talked with Harold ("Chuck") Klein, director of life sciences at NASA's Ames Research Center in Mountain View, California, who led the biology team for the Viking mission. In 1975 Dr. Klein commented that he was "pessimistic about the reality of . . . life on Mars, but open minded because we have been surprised so much by Mars. The chances are low; 1 in 50 to 1 in 500." Several years earlier when I had met him at an international meeting in Seattle, he had pointed out, however, that if "the data [from a landing on Mars] indicate a planet devoid of organic matter or of living organisms, this would lead to a new speculation and theories about the origin and development of our solar system."

The position had completely reversed from 100 years ago. Then most scientists would have said that Mars was the abode of life of some kind, even if it was not intelligent life. Just before the two Viking spacecraft were launched in the summer of 1975, most scientists doubted that there was life on Mars. By contrast, the public expressed more generally the opinion that the Vikings might find life.

Undoubtedly Mars still fascinated modern man; Martians seemed real to people who had grown up in the age of science fiction and novels such as those of Edgar Rice Burroughs. As the launch date for the Viking spacecraft approached, people worldwide began to ask about the mission to Mars. News media representatives made pilgrimages to NASA's Langley Research Center in Hampton, Virginia, from which the Viking project was managed; to the Jet Propulsion Laboratory in Pasadena, from which the mission operations would be directed; to the Ames Research Center, which had responsibility for the biology experiments on the spacecraft; to the contractor for these experiments, TRW Systems, in Redondo Beach, California; and to the Martin Marietta Corporation in Denver, the principal contractor for the spacecraft and the Viking system.

Many people began asking the questions, Life on Mars? What if? The words of Dr. Horowitz after the Mariner 6 flyby of Mars in 1969 seemed most applicable: "The discovery of life on another planet is very important to man. As long as we have the technology to get answers, we are not only justified but obligated to search." And from Dr. Klein: "Even those of us who are on the pessimistic side are absolutely convinced that the payoff if we are wrong, that is, if there is life [on Mars[, is enormous."

Mars was only a red starlike object in the night sky to a breed of primates that lived on the earth for several million years. Then it became an indistinct globe of a

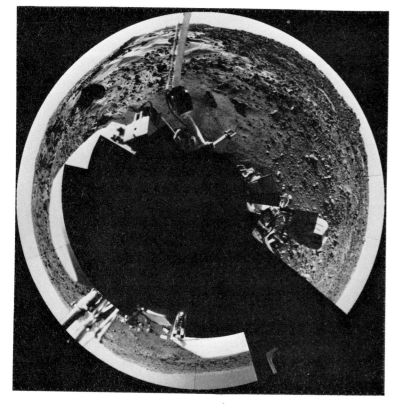

1.4 This "fisheye" view from Viking 1 shows a view of Chryse Planitia around the landed spacecraft. It was created in a computer by geometrically transforming mosaics of images of the Martian surface obtained by one of the cameras on the first American spacecraft to land on Mars.

mysterious world to those astronomers who looked at it through telescopes for 300 years. But in the 40 years that I had been interested in Mars it had become a place, or rather a series of places on a complex world that was neither earthlike nor moonlike nor Venus-like, but a world of its own. And through the special cameras of a spacecraft I was to see the surface of Mars as clearly as I see the surface of the earth when I stand upon it and look around me (figure 1.4).

In this I consider myself a part of the unique generation of humans alluded to by Carl Sagan in a talk about Mariner 9 results. Dr. Sagan and I were among a score of speakers talking to passengers of the S.S. *Statendam* about the voyage beyond Apollo on the occasion of the Apollo 17 launching in December 1972. Dr. Sagan, an astronomer who had long advocated strongly the search for extraterrestrial life, was a prominent member of the science teams connected

with the Mariner program to photograph Mars from spacecraft that flew by and orbited the planet. He pointed out that before our time no one knew anything really definite about Mars. After our time no one can suspend disbelief to enjoy the entertaining pre-spaceflight speculations. "There is only [this] one generation in the whole history of mankind," said Sagan, "all generations earlier wondered and never found out; all generations later have found out and never wondered." We are unique for we have lived during the transition from interplanetary ignorance to interplanetary knowledge, a transition that covered other worlds of the solar system as well as Mars.

This knowledge about Mars cost me as a taxpayer a paltry five dollars, plus the same amount for my wife and three children—a total of twenty-five dollars—to find out what Mars is like. This amounts to only one dollar a year spread over the 25 years since my first technical article on a plan to send a probe to Mars. I wish everything I have been curious about during my life could have been resolved for so little.

In the context of history, it seems certain that Viking's landing on Mars will be remembered long after the energy crises, the political turmoils, and the recessions of this age are forgotten; just as the rediscovery of America by Columbus stands out while the misery, disease, and squalor of medieval Europe are largely ignored in the history books.

Humankind does not remember the frustrations and the failures. History rewards the successes with a place in the human archives, and Viking on Mars has been one of the major successes of this decade.

Mars in the Human Mind

Chapter 2

Life on Mars: What If?

The question of what is life has continued to intrigue philosophers, theologians, natural scientists, and lay people alike. Leonardo da Vinci is said to have defined life as expressing the power to move about. In modern times scientists have defined it as organization of inanimate matter into a system that perceives, reacts to, and evolves to cope with changes to the physical environment that threaten to destroy its organization.

Attempts have been made to define life in terms of ability to reproduce, to store information, to adapt genetically, to react to and protect itself against a changing environment, and to change the chaos of the physical world around its system into a planned order within its system. But these definitions are based on observation of terrestrial life forms that share the same fundamentals, as though all life on earth arose from only one source, as it may well have.

A major problem in searching for extraterrestrial life is that no definition of life is generally accepted. Although astronomers have shown that general laws of physics and chemistry extend to the distant nebulae, generalizations of biology cannot be extended beyond the earth because no evidence yet exists of life elsewhere in the universe.

In searching for life beyond the Earth, we can make either of two assumptions: that any life elsewhere is based on the same carbon chemistry as terrestrial life or that it is based on some completely different chemistry. Terrestrial life is structured from elements that are particularly abundant in the universe: carbon, nitrogen, oxygen, hydrogen, and traces of at least 25 other elements. In the human body, as in the stars, hydrogen is the most abundant element. Oxygen is next, and then carbon. But carbon plays a very important part because it is so versatile; it can take part in many types of chemical reactions and is used to make complex biological molecules. Some science fiction writers have suggested that other life forms might be based on silicon instead of carbon, but scientists believe this is unlikely. Both carbon and silicon can form four covalent bonds. However, there is a major difference. An atom of carbon coupled by these bonds to two atoms of oxygen forms the extremely stable carbon dioxide, a gas that is readily soluble in water and always remains in single molecules. By contrast, silicon dioxide clumps into large molecules with very different characteristics from carbon dioxide; its most familiar form is sand. Moreover, silicon atoms do not form long chains as carbon atoms do. Carbon atoms have an almost unique ability to link with one another in these chains, in stable rings, and in other structures, to which other elements are also attached. Silicon chains are formed by alternating bonds with oxygen (producing the silicones), which are difficult to break as compared with carbon chains.

Earthly life forms build up, or synthesize, complex molecules based on carbon (known as organic molecules) in a process which chemists term reduction; they break down these molecules by a process of oxidation to release energy. Carbon can readily undergo both oxidation and reduction reactions, thus making it possible for electrons to flow either way. In photosynthesis, for example, captured solar energy adds electrons to carbon and reduces it to organic matter. Respiration or fermentation takes electrons from the organic compounds and oxidizes the carbon. Silicon as an element cannot easily function in this dual manner.

Most scientists opted to search for extraterrestrial life forms that are based on carbon chemistry. Since such life forms may be at different stages of evolution, tracing the genesis of terrestrial life is important in designing experiments to look for life on other worlds.

In recent decades scientists have sought answers to how and when living things first evolved on earth from inorganic molecules. In the early 1800s geologists established relationships between rock forms and the types of fossils found within them—fossils from higher rock layers more closely resembled modern plants and animals than did those from deeper and presumably older rocks. A relative time scale was established for the rock strata. Later, when some igneous rocks could be dated by radioactive methods, an absolute time scale was established for the rocks, thereby setting a sequence through Cenozoic, Mesozoic, and Paleozoic eras for some 600 million years.

Before 1950 most scientists believed there were no fossils in rocks of the earlier Precambrian era. However, in 1953 an economic geologist, Stanley Tyler, who was unaware of the dogma then prevalent in scientific circles, looked for fossils in the Gunflint Cherts on the north shore of Lake Superior. He found some that were 2 billion years old. Many older fossils were later found in Precambrian era rocks—in the Fig Tree and Bulawayan rocks of South Africa, and at Bitter Springs in Australia.

My first acquaintance with paleobiologists took place in May 1969. I was in Tucson to research a story about growing food crops under arid conditions using seawater for irrigation. I found that there was a colloquium on organic geochemistry at the University of Arizona and I decided to look in on it. There I met Bartholomew Nagy, a professor of geochronology, and I listened to Nobel Laureate Harold Urey, and William Schopf of the University of California, Richard S. Young of NASA, and John Oro of the University of Hawaii, discussing the question of how life on earth might have evolved from prebiological molecules and what evidence there was for extraterrestrial life.

Dr. Nagy described his work with some of the oldest rocks on earth, such as those overlying the Onverwacht sediments of Swaziland, in Africa. The Fig Tree sediments were deposited about 3 billion years ago, the Bulawayan about 2.8 billion years ago, and the Gunflint about 2 billion years ago. By pulverizing samples from these deposits and looking at them with an electron microscope, he had discovered tiny rodlike shells and spheroids that might be interpreted as very early forms of terrestrial life. The evidence suggested that life on earth consisted of simple cells about 3 billion years ago. It seems that somewhere between 3.4 and 3.0 billion years ago nonbiological molecules became living systems able to replicate themselves, and this was the beginning of life on earth. From about 3.2 billion years to the present, scientists can trace a continuing evolution of these living systems.

Life on Mars: What If?

Unknown to each other, the Russian Alexander Oparin and the Englishman J. B. S. Haldane proposed in the 1920s that life evolved spontaneously on earth because of peculiar chemical and geological conditions here. Their detailed papers were not published until the 1930s. Oparin's original 1924 paper suggested that hydrogen-rich substances such as methane, ammonia, and water were essential to the formation of the first living cells. Independently, Haldane suggested in 1927 that life had developed deep in a primitive ocean that was rich in organic compounds built around a carbon base.

Stemming from these early theories, other scientists developed a four-stage sequence for evolution of life on a planet: nuclear evolution, chemical evolution, transition to living things, and biological evolution.

NUCLEAR EVOLUTION. During this first stage, hydrogen within stars was transformed into heavier elements such as carbon, nitrogen, oxygen, sulfur, and phosphorus. Some stars exploded and scattered these heavier elements through the galaxy, thus making them available for the next evolutionary stage on cooler bodies than the stars.

CHEMICAL EVOLUTION. During this next stage, hydrogen atoms linked to the heavier elements to form the molecules that are so much a part of life on earth today, such as water, methyl groups of hydrogen and carbon, and derivatives of ammonia containing hydrogen and nitrogen. These are sometimes referred to as biomolecules, and they have been identified in interstellar dust clouds and in meteorites. During the 1930s simple molecules made of carbon and hydrogen and carbon and nitrogen were detected between the stars by optical methods using the 100-inch (250-cm) Mount Wilson telescope. But it was only in 1955 that Charles Townes of the University of California suggested using radio telescopes to look for more complex interstellar molecules. Since most scientists believed that ultraviolet radiation from stars would prevent the formation of or would break down any polyatomic molecules in interstellar space, only the simplest molecules were searched for.

In 1968, radio astronomers from the University of California at Berkeley, using a 20-foot (6-m) diameter radio telescope at Hat Creek (near Mount Lassen), detected ammonia in interstellar clouds. That same year they also detected water. In succeeding years more powerful instruments and improved observing techniques allowed the detection of many complex molecules. Interstellar space began to look like a gigantic laboratory for organic chemistry. Scientists became even more excited when most of the molecules discovered in interstellar space were the ones that are important in natural reactions to synthesize amino acids, which in turn are utilized by terrestrial life forms to build proteins.

Some people speculated that such biomolecules might have been present in planets when they originally formed from the dust and gas of an interstellar cloud. But it is difficult to imagine how these molecules could survive when the planets later melted to form cores and crusts. Others speculated that the biomolecules might arrive on a planet after it had formed a crust, perhaps carried by meteorites or comets. And experiments by biologists showed that amino acids could originate on a planet spontaneously.

At the 1969 Tucson colloquium mentioned earlier, Dr. Urey warned that we had no strong evidence that meteorites contained organic material. The meteorites had not been analyzed soon enough after they fell to earth and had probably been contaminated by terrestrial life. He mentioned a possible exception. This was a carbonaceous chondrite which fell on February 8, 1969 in Mexico. (This class of crumbly meteorites is thought to be representative of the materials from which the inner planets were formed.) The meteorite made a brilliant fireball over southern

Arizona, and about 1 ton of fragments was recovered on the ground near Pueblito de Allende. Some of the fragments were rushed to the Jet Propulsion Laboratory for detailed chemical analysis, and organic molecules were found in them.

Speaking at the fourth international symposium on bioastronautics at San Antonio, Texas, in 1968, Cyril Ponnamperuma of NASA's Ames Research Center had stated: "Laboratory experiments have shown that, wherever suitable conditions exist, organic compounds of biological significance can be synthesized. These results lend support to the hypothesis of chemical evolution and to our belief in the existence of extraterrestrial life."

I talked with Dr. Ponnamperuma in November 1970, soon after the announcement that he had found amino acids in another meteorite. This one flashed through the skies of Australia on September 28, 1969 and landed near Murchison, Victoria. It was found and rushed to Ames before there was much chance of contamination by terrestrial materials. Using the most advanced tools for biochemical analysis, a team headed by Dr. Ponnamperuma discovered that the meteorite contained amino acids.

Although contamination could not be ruled out entirely, a new approach was taken to identify the biological materials in this meteorite. Instead of searching for products of the biological process—for evidence of organic life—the NASA researchers looked for amino acids and hydrocarbons, the chemical building blocks of our kind of life. Then they sought to identify differences between these molecules in the meteorite and biological building blocks common on earth. They were successful. Five of the 20 amino acids normally present in terrestrial living things were discovered in the meteorite. But these could very well have come from the earth. However, the meteorite also contained 11 other amino acids that

are not present in terrestrial living things. Hence Dr. Ponnamperuma concluded that the meteorite's amino acids were extraterrestrial. Their presence showed that amino acids could be formed from more simple molecules without preexisting life. "The find is probably the first conclusive proof of extraterrestrial chemical evolution—the chemical processes which preceded the origin of life," he said.

His conclusion was supported by other evidence from this meteorite. The amino acids have special molecular arrangements not common to earth life; if earth amino acids are compared to a left-handed glove, the space amino acids were like right-handed gloves. And there was also a mixture of biological and nonbiological hydrocarbons in which a heavy isotope of carbon was present in greater proportions than found in earth's biological hydrocarbons. Because the meteorite was dated as 4.5 billion years old, it seemed to point to a general availability of complex life materials from the time of the earth's formation.

The Murchison meteorite was believed to originate in the asteroid belt between Mars and Jupiter, and it added considerable credence to speculation that Mars, too, must have had prebiological organic compounds on its surface early in its history. Once such molecules are established on a planet, they could next evolve into prebiotic amino acid chains and other more complex molecules that have been called building blocks of organic matter. Experiments in laboratories have shown that heat, ultraviolet light, or electrical discharges can provide the energy to synthesize the prebiological molecules in a chemical evolution from simpler molecules; the discoverer was Stanley Miller.

Dr. Jerzy Neyman, who is often referred to as the father of modern statistics, edited *The Heritage of Copernicus*. When I visited him at the University of California in Berkeley to discuss the revolution in thought that might come from the discovery of life on Mars, he described why he believed Dr. Miller had made a significant contribution to the Copernican revolution. In the early 1950s most scientists believed that the early atmosphere of the earth did not contain oxygen and nitrogen, but was mainly methane and ammonia. (It is now thought to have been carbon dioxide, carbon monoxide, nitrogen, hydrogen, and water vapor.) During a 1951 seminar at the University of Chicago, Harold Urey talked about the possibility of synthesizing organic macromolecules by lightning. Stanley Miller was one of his students. He told Urey he would like to experiment with synthesizing organic compounds in an atmosphere like that assumed to be the atmosphere of the primitive earth. Miller suggested it could be the subject of his thesis.

Dr. Urey was reluctant at first because such a project had many elements of risk and might not be completed within the two to four years normally expected for a Ph.D. thesis. But he gave Miller a conditional go-ahead for a year. By the end of 1952 Miller had produced a number of organic compounds, including three different amino acids, in his laboratory experiments and was ready to publish his exciting results. They appeared in *Science* on May 15, 1953. Since then, similar experiments have been successfully carried out by many other chemists, modifying the mixture of gases to accord with later theories of the primitive atmosphere and using ultraviolet radiation as the source of energy. Natural processes seem to be capable of evolving carbon compounds ready for a transition into living things.

TRANSITION TO LIVING THINGS. How the third evolutionary stage of transition took place is still unknown. There is no evidence in the fossil record of earth to show exactly when or how prebiological molecules changed to biological ones. Nor has a process for this transition been demonstrated in any laboratory, though Sidney Fox showed in the late 1950s that dry amino acids could be converted by moderate heat into chains of polypeptides which he called "protenoids." Some even possessed catalytic properties, like rudimentary enzymes. Immersed later in water, they formed spheres with membranes, simulating the form of the earliest living cells.

In recent years scientists have performed many chemical evolution experiments and produced most of the basic life molecules, including amino acids and nucleotides, but they could not explain how these building blocks were collected and organized into living cells. Only in 1977 did scientists at the Ames Research Center find that metal-clays can concentrate amino acids into proteins and DNA chains. James Lawless led a team of scientists that discovered how building blocks of life may have been collected and organized on the shores of primordial oceans by natural catalysts of metal-clays. A metal-clay containing nickel concentrates the 20 amino acids which make protein, the main structural ingredient of living cells. Experiments simulating tidal action in drying, warming, and rewetting an amino acid–clay solution produced chains of amino acids that could be the first steps in producing the longer chains found in living systems. The experimenters also discovered that a metal-clay containing zinc had a similar effect on nucleotides, which are the building blocks for the complex molecules that contain the genetic blueprint for each living organism.

Once the self-replicating system of a living cell had been achieved, the fourth stage of continued biological evolution produced more complex forms that led to the many varied types of living systems we have on earth today. Life may thus have originated on earth from protenoids produced on hot dry rocks and washed into the primitive ocean, which provided conditions in which life learned to reproduce itself by cell division.

BIOLOGICAL EVOLUTION. The elementary particle of terrestrial life seems to be the biological cell, a small, self-contained chemical factory that both builds up and breaks down complex molecules of carbon compounds. Within a range of relatively low temperatures these large molecules are stable. Yet they can also easily participate in chemical reactions under the influence of catalysts, which speed up chemical reactions without being used up by the reactions. The biological catalysts are called enzymes.

The simplest living organism is a single cell. Higher organisms consist of millions, or billions, of cells organized into a cooperative whole. Even the single cell is a maze of complexity. Its essential features are a bounding membrane, a source of energy, catalysts, a capacity for regulation and repair, an inherited information and control center, and a means for nearly precise reproduction of the store of inherited information that is transmitted from generation to generation. All cells of all life on earth come from preexisting cells. So far as we are aware, no living cell has been spontaneously generated. However, the recent discovery of methane-producing organisms in hot springs has led to speculation about a separate genesis of this life form and those of bacteria and the animal–plant life forms of the earth.

Each cell has a bounding membrane to protect it from the surrounding environment and from potentially harmful agents, and to prevent its specialized components from diffusing into the environment. This flexible, self-made, and continuously repaired membrane forms the boundary to the cell's living organized world. Under control from within the cell, this membrane determines what enters and leaves the cell. The membrane consists of two stacked layers of lipid (fat-related) molecules with protein molecules to act like revolving doors to transport substances through the membrane.

Cells need energy to function. The first living organisms on earth were single-celled creatures that probably depended on the environment for previously formed organic molecules (from chemical evolution) which they ingested and used to derive energy by rudimentary enzyme activities. But when these organic molecules were nearly used up, the early living things had to find energy elsewhere. Some of these cells adapted. They made their own food by developing an ability to use sunlight to fix the carbon dioxide and nitrogen of the atmosphere to create their own organic reserves, and released oxygen as a by-product into the earth's atmosphere. The blue-green algae used sunlight to separate water into hydrogen (which they burned for energy) and oxygen. They also used nitrogen from the air to make amino acids. Since they lived independently on only air, water, light, and a few minerals, they were called autotrophs ("self-nourishing"). The blue-greens still exist, almost unchanged, and their evolutionary descendants, the green plants, have taken myriad forms.

Later other forms of terrestrial life evolved that were capable of "eating" the autotrophs and using them as a source of energy. The newer life forms used the atmospheric oxygen in energy-releasing metabolism. In the process, these "heterotrophs" recycled some of the organic material back to carbon dioxide. Thus the two types of organisms—whose highest forms are plants and animals—together completed a biological cycle that continues to the present. In the photosynthesizing life forms, specialized groups of molecules called chloroplasts convert light energy to chemical energy. In the heterotrophs, cell structures called mitochondria provide chemical energy from the controlled breaking down of chemical bonds (oxidation) of complex nutrient molecules, producing carbon dioxide and water.

Cells use enzymes to increase the rate of desired chemical reactions up to 10,000 times faster than these rates would normally be. In this way the chemical reactions needed for the process of living are caused to proceed in the specific direction required.

Another crucial feature of a cell is the capacity for regulation and repair. Although living systems appear to be able to maintain organization and order in the face of innumerable other forces seemingly tending toward disintegration, the essence of life does not seem to be solely organization. Any complex chemical organization would inevitably disintegrate. An important factor in survival is the ability of living things to repair themselves to prevent disintegration. Each cell makes all its component parts and maintains and repairs them. It possesses a management system to keep its many variables regulated and in balance for the tasks that the cell must perform as a whole and its many parts must perform individually to retain the organization.

Thus, the cell requires an inherited biological information and control center consisting of genetic material that contains molecular blueprints to specify, in a language common to every form of life on earth, the structure of all cellular components and all the catalysts needed for the cell's chemical reactions. Life on earth combines nucleic acids and proteins; the former are the self-replicating components that carry the genetic message, the latter direct the complex chemistry of the living organism (many are enzymes). These two components interlock in the living system.

The simple cells of the first living microbes dispersed their genetic material within their cells. The next important milestone in biological evolution was development of the eukaryotic cell, a cell with a nucleus in which the genetic material was concentrated. Ultimately this development led to sexual reproduction, in which a recombination of genes from two parents made possible great variety and the ability to meet changing environmental conditions. It allowed a fantastic diversification of living forms, an almost unlimited biological experimentation that led to higher forms of plants and animals. Recently it produced technological man and his ability to search for life on other worlds.

Instructions for assembling protein molecules that make up the structure of a cell are encoded in nucleic acids in sequences of four different nucleotides. These are molecules consisting of a pentose sugar to which is attached both a phosphate group and one of four bases. In ribonucleic acids (RNA), the bases are adenine, guanine, cytosine, and uracil; in deoxyribonucleic acids (DNA), thymine replaces uracil. The DNA of the genes controls the formation of RNA molecules, which in turn control the assembling of cellular proteins. The only genetic differences among life forms on the planet are in the linear arrangement and the amount of nucleotides that make up the genetic message for each of them. The differences among the many proteins used to structure and operate the living system are in the arrangement of about 20 amino acids. The nucleus of every human cell, for example, carries some 5 billion bits of information in its genetic message. But it is the instructions that are copied, not the relatively bulky sequence of amino acids. The vital molecular message to do this consists of double-stranded helix molecules of DNA, in which one strand is the exact complement of the other. The complementarity is possible because an adenine on one strand normally will match up only with a thymine on the other, and guanine and cytosine are similarly paired. A sequence of ACT on one strand will be matched by TGA on the other. The message is preserved over eons of time by means of redundancy and repair. Molecular redundancy in this double-stranded DNA recognizes any damage to one of the strands and repairs it from the template of the other. This molecular helix can replicate itself very accurately, making only one mistake in about 100 million replications.

Although the theory of evolution of life through nuclear, chemical, transitional, and biological stages seems attractive and logical, there is no way of checking its validity (especially for the chemical and transitional stages) without finding life elsewhere than on earth. As a result, the 1950s and 1960s saw the development of a plan to start searching. The development was encouraged by the discovery that pre-biological molecules appear to be common in the universe and that terrestrial life forms are extremely tough and can live under a wide range of environmental conditions. Many can thrive in what might appear to be lethal environments.

Stanford Siegel, a professor of botany at the University of Hawaii, had already spent many years investigating limits to life when I met him in 1970. Dr. Siegel was reshaping theories of organic life in his laboratory, continuing work he had started in London. He placed tarantulas in an environment simulating air pressure at 8 miles (12.8 km), and they remained active and aggressive. When they were exposed to ultraviolet radiation for a month, they still remained apparently unharmed. Baby turtles lived in an atmosphere of toxic carbon monoxide. Plants and bacteria proved even more hardy than animals. He took a cactus, evolved for the desert, and grew it underwater in an aquarium. He had found bacteria, probably descendants of survivors from the initial forms of earth-life, which thrived in an atmosphere of ammonia. Today such bacteria bury themselves deep in soil to avoid oxygen, which is poisonous to them.

Survival is not enough, however. Organisms must be active and reproduce. Seeds or spores are resting stages at which organisms are extremely resistant to heat and radiation. It may be that feedback from the environment as a life form evolves decides the pattern of the full-grown organism. For example, when Dr. Siegel sprouted seeds without oxygen the adult plants could not tolerate an oxygen-rich atmosphere.

Conversely, seeds sprouted in oxygen could not tolerate, as plants, an oxygen-deficient atmosphere. Even more unusual, mold spores that were kept in liquid ammonia at −40 degrees Fahrenheit (−40°C) for months became accustomed to the environment and began to make nucleic acids and proteins and started to grow.

At low temperatures it is not so much the cold that prevents life from thriving, as the change of water from a liquid to a crystalline state. (At the high end of the temperature scale, the change of water to vapor and the coagulation of proteins cause problems.) That is why some living things can occupy ultra-cool ponds of brine at a temperature below the normal freezing point of fresh water. Experience of the tenacity and adaptability of life on earth showed that living things might be able to adapt to and even flourish in the inhospitable environment of Mars.

The world's first institute for the study of extraterrestrial biology was formed at Alma-Ata in Kazakhstan in the 1950s, associated with the Alma-Ata Observatory. It was called the Institute for Astrobiology. Even earlier, in 1947, G. Tikhov (a Soviet astronomer who began his studies of Mars at about the time when Lowell published his controversial book *Mars as the Abode of Life*) had written a book on astrobotany. He wrote another on astrobiology in 1953.

Hubertus Strughold is another founder of extraterrestrial biology and space medicine. In 1954 he published *The Green and Red Planet,* which attempted a thorough scientific analysis of the possibilities of life on Mars and concluded that Mars was not really suited for life as we know it. I was a little disappointed at the conclusion, but I still liked the book. During the 1960s Dr. Strughold organized several international symposia on bioastronautics, at which preliminary ideas for searching for life on Mars were aired. Despite his conclusion about possibilities of life on Mars, he obviously thought that the search was worthwhile.

"Extraterrestrial biology" is something of a tongue-twister as a name for the science of biology beyond the earth. Joshua Lederberg, a Nobel laureate geneticist at Stanford Medical School, suggested the term exobiology, and it has become generally accepted. Dr. Lederberg began his active work in this field in 1959, when Stanford received a small grant from the Rockefeller Foundation to lay groundwork for experiments to find out whether life has evolved on other planets. The outcome of this work and that of other scientists who entered the new field of exobiology is described in the next chapter.

Meanwhile, planetologists were developing a new view of Mars as a possible abode of life. So far as we know, three things govern whether or not a planet can develop and support life. First, there must be an atmosphere. Then there must be water. Third, the range of temperatures on the planet must be right. Mars certainly has an atmosphere, it does have water, and its surface temperature is within a range that is marginal for some forms of terrestrial life.

Early in the development of programs to search for life on other planets, exobiologists warned that one simple experiment could never cover all possible manifestations of life. No detector could be devised to give a clearcut answer that life exists on another world, unless it is an easily recognizable creature like a large terrestrial life form. Microbes are very difficult to detect. In searching for living things on other worlds, we might look for their properties and the way they react with inanimate matter—such as metabolic activity, growth, capture and use of solar energy in photosynthesis, absorption of other life forms in an eating process, cyclic oxidation and reduction activities, respiration, and so on. As the programs developed, scientists concentrated on developing instruments that could look for three aspects of living things as we know them: manufacture of food (like plants), consumption of food (like animals), and changes in atmospheric gases (like both plants and animals). The development of these instruments is described in a later chapter.

The Viking expedition rekindled an excitement experienced in earlier centuries of exploration on earth; in looking for life on Mars, we had to expect the unusual and perhaps the bizarre. Biologists connected with Viking experiments thought their chances of finding life on Mars were very small, but they admitted they could be wrong. Finding a single example of any extraterrestrial life form would have tremendous implications for our understanding of terrestrial biology and the origin of life. One of the central things that pushes science on to new discoveries is encountering the unexpected and then finding explanations for it. Scientists hoped that if extraterrestrial life were found they would then be able to learn whether it, too, was based on carbon and oxygen and photosynthesis, or differed in central features of biochemistry or included a hereditary apparatus that was fundamentally different from terrestrial life. If a Martian life form were discovered and it could be shown to have developed spontaneously on Mars and evolved in the same way that terrestrial organisms had evolved, this would go a long way toward showing that extraterrestrial life might generally evolve into intelligent creatures.

Suppose we could find out whether Mars harbored life—what might this mean? Various philosophers, theologians, anthropologists, and other people had differing views when I asked what they thought about this search for extraterrestrial life and its consequences to human thought and to society.

An astronomer who concentrates on distant star systems rather than nearby planets asked me, when I interviewed him, "What is . . . Martian life? Will we find bacteria, spores of some simple fungus, some very elementary single-celled life, or something much more complicated, as complex as an insect? Will we find something that we cannot describe as animal or plant? Suppose we were to find ruins eroded by the sandstorms of Mars; how exciting that would be! We don't think that there are higher life forms on Mars now, but there could have been."

Would there be panic, as in *The War of the Worlds,* if we discovered a race of Martians? I asked a well-known futurist and social anthropologist who had recently helped an African people adjust to displacement when a new dam flooded their homeland. He replied that "on Earth natural disasters produce a characteristic behavior pattern. People behave as if the social system is closed. They fragment into small groups, families, or neighborhoods and stick to familiar people or strategies to cope with the disaster. They are not willing to accept radically different ideas from within or without because that only increases the stress. Meanwhile, outside communities rally and come to the rescue and return things to normal." An extraterrestrial threat might produce the same kind of response—no panic, but fragmentation. Natural disasters are usually localized. An extraterrestrial threat could be global; not necessarily intelligent invaders, but microbial contamination of the earth. If the threat were localized, a logical response would be reinforced from the rest of the world. But if it were global, with the whole world threatened, there would be no outside relief and there could be an extraordinary degree of fragmentation, with every community acting as though it were a closed system.

What effects would the discovery of Martian life have among people in less developed regions of the world? The belief systems of many peoples do not make a rigid distinction between the natural and the supernatural. The prevalence of animism, in which spirits are associated with trees and rocks, and the fact that

the stars and moon play a major part in their agriculture lead to the assumption that people in many less-developed areas would expect to find forces which they would define as life "out there." The impact would be different for different world views, but as a general proposition such people might find the whole idea of extraterrestrial life more congenial than would people in Western civilization. This follows from their relationship to nature and the supernatural.

Politically and religiously, however, there are problems of encountering extraterrestrial life, especially if it should prove to be intelligent. Russian and Chinese communism from the ideological standpoint, and Judeo-Christian theologies from the religious standpoint, claim a missionary primacy, making members the center of the universe. Such egocentric world views may not be able to accept other world intelligences greater than their own.

On this question of religious acceptance, an authority of world religions at a major university doubted whether the discovery of life on Mars, or the discovery that there is no life on Mars, would have an immediate major impact: "In the short term the reaction will be to fit the discovery into the existing framework, saying that such and such a statement in the religious writing shows that the result was already forecast. But in the long term it must affect the adjustment in perspective that the world's religions are facing today."

Suppose there is no life on Mars, yet all conditions on the planet appear suitable for life to have developed, even if not to our advanced form. This finding might reinforce the belief that a single creative event in our Milky Way galaxy took place on earth and the thought that humans are alone in the galaxy with a "torch" to carry. (The fundamentalists would feel vindicated.) Then mankind could have a new cosmic consciousness of our true destiny, which could strongly influence all activities here on earth. The purpose of mankind might then be identified as cosmic diffusion, analogous to the planetary diffusion from an origin in Africa.

I questioned educated laymen too. I talked with a businessman who had experienced the horrors of a Nazi concentration camp and had seen the worst as well as the best of human life. His view was that "anything less than seeing any live creature . . . would be a disappointment, because we always think there must be something out there. But with disappointing results it is not necessary to believe that we have reached the ultimate, that there is no other life. I'll say this, if there is other life it will be of a completely different form." He was enthusiastic about the search: "We have to do everything; we can't give up the search for more life and knowledge. It is not a matter of money; but is a matter of new horizons, new hopes, and new possibilities. Without these, people will stagnate."

Another typical comment was: "My reaction to the discovery of life on Mars would depend very much on the nature of the discovery. Is it like us, or is it a bug-eyed monster?" Life there might be clearly related to life on earth, with the same amino acids, the same proteins, the same genetic code structure. "If we find this on Mars, what a train of possibilities it would open up!" exclaimed one woman. It might, indeed, show that life forms evolve toward a common end irrespective of the world on which they develop.

How about intelligent life? Intelligence seems to have developed on earth because of a need to adapt to a rapidly changing and complex environment. Perhaps on a planet with a constant climate intelligence would not develop. The intelligence of whales and dolphins, for example, seems older than human intelligence but has not evolved as far or as rapidly. In the benign marine environment, whales and dolphins have had no need for manipulative skills and so did not progress to technology. Might the inhospitable

environment of Mars have forced life there into manipulative intelligence?

On the other hand, Martian life might be completely different from terrestrial life. The part of our evolution that seems most difficult to duplicate is not the original emergence from prebiological molecules, but the development of the cell with a nucleus which led, in turn, to multicellular organisms with sexual reproduction and to mankind. This capability of storing genetic information in a cell's central necleus, of developing an integrated mass of specialized cells of different kinds, all in communication with one another, and increasing the amount of variety in offspring through sexual reproduction, made rapid evolution possible.

Moreover, suggested a sociologist, the discovery of living systems quite different from those on earth could push us into rethinking what the nature of society is and what its social institutions should be. The study of animal societies, of chimpanzees, for example, drastically changed man's picture of himself in relation to other animals. Today, as a result, it seems that the gap between man and the animals does not require any unusual evolution that could not occur on another planet. Once a life form as complex as the mammals had developed, it might always, given time, evolve high intelligence.

While the analysis of meteorites had been encouraging to scientists looking for evidence of extraterrestrial life, as described earlier, the results from a search of lunar rocks were not. Amino acids were not found in the samples from the moon; in fact, no complex carbon compounds of lunar origin were found. Despite 3,000 tests for microorganisms, all results were negative, even though some terrestrial microorganisms did survive within parts of a Surveyor spacecraft and were found when these parts were later returned to earth by Apollo astronauts. The lunar samples indicated a moon devoid of living things

Life on Mars: What If?

since its formation. There were no microfossils, there was nothing that could be interpreted as biological in origin, and no single organic compound that might be significant to biology was found.

The moon had been the first testing ground for the theory that life may be disseminated through space from world to world or might originate spontaneously on all planets. And the results had been negative. Would the planet Mars prove equally disappointing? Or did Mars have life forms that had developed in the distant past, when the planet flowed with liquid water, and were now adapted to dryness far exceeding that of the most inhospitable terrestrial deserts? The Viking mission to Mars might answer such questions.

Should the mission to Mars fail to show any proof of life there, this would not alter current scientific thinking about life elsewhere in the universe. But it is important to understand that a negative result from Viking would not *prove* that no life exists on Mars. It would only show that where the spacecraft landed there was no evidence of life of the type we are looking for, i.e., microbial life that makes its own food from the Martian atmosphere of carbon dioxide, or consumes food brought to it from earth, or interacts with its gaseous environment, or large life forms visible in pictures showing living things or their artifacts.

We do know that life on a planet can be very important to the planet's evolution and even to the development of a technological society. Our technology has been strongly based upon the use of iron and steel. Yet the iron ores we have mined to build our machines and our bridges were made possible by the pollution of a lowly plantlike creature over 2 billion years ago. These early blue-green algae started to make oxygen as by-products of photosynthesis and within a few million years polluted the primitive at-

mosphere of earth with this gas, which was poisonous to many other life forms that could not adapt to protect themselves against it. It also trapped much of the earth's surface iron into oxides that became great deposits of iron ore that spawned a technological civilization two eons later. And the "poisonous" atmosphere led to the development of mammals and man because it allowed a much more efficient energy metabolism for cells.

The important aspect of the Viking mission was that it was not going to the Red Planet only to look for life there; it was basically designed to find out more about another planet through observations made on its surface. Looking for life was only one aspect of these observations. Viking would not be a failure if it did not find life on Mars. In science an experiment is still a success even if its results provide a resounding no.

Expectations make a big difference, however, in the human mind. A psychologist answered my question by saying, "If people believe strongly in there being life on Mars, they will be disappointed if none is found by Viking. But if they do not expect life on Mars in the first instance, they will not be surprised either way."

Still, as they started to think more deeply about the possibility of a spacecraft landing on Mars, many people experienced an anticipatory excitement perhaps best defined in a frequent comment: "We must be prepared for surprises." In 1492 Columbus discovered the New World and new people. Within 100 years afterward, mankind also benefited from the Copernican revolution, the Reformation, the Renaissance. The answer to the question of "life on Mars, what if?" held promise of coloring mankind's future and governing the attitudes of people toward their own world and their own destiny.

Life on Mars: What If?

Chapter 3

The People and the Plan

The year 1971 was important for our understanding of Mars: It was a good year to see Mars from the earth, the year of an unusually good opposition (when the earth is directly between the sun and Mars). Two NASA spacecraft were to be launched to orbit Mars. And the NASA Viking project to land two spacecraft on Mars became fully active again after a cutback in activities because of a shortage of funds.

When Mars is in opposition it is also closest to the earth and shines its brightest in the southern sky at midnight (in the northern sky from the southern hemisphere). Oppositions of Mars take place every 26 months on the average. But because Mars' path around the sun is an ellipse rather than a circle, the planet moves between two extremes of distance. At its closest approach to the sun, which is called perihelion, Mars is 128.42 million miles (206.66 million km) from the sun. At the most distant part of its orbit, called aphelion, Mars is 154.86 million miles (249.22 million km) from the sun. Whereas the distance between the sun and Mars thus varies by 26.44 million miles (42.56 million km), the distance between the sun and the earth, which has a more circular orbit, varies by only 3.22 million miles (5.2 million km).

Some oppositions occur when Mars is near its perihelion; a perihelic opposition brings Mars closer to the earth. These oppositions occur about 16 years apart, but not all are equally close because they do not occur at precisely the perihelion point. Very good oppositions take place about twice a century. The first in this century occurred in 1924, when Mars was only 34.66 million miles (55.78 million km) from earth on August 22. At this favorable opposition the canal controversy was revived.

Robert Trumpler reported on photographic and visual observations of Mars made during the 1924 opposition period with the 36-inch (91.4 cm) refractor at Lick Observatory under excellent seeing conditions. Before this opposition he had not systematically observed Mars and was not familiar with Martian topography, but he produced a detailed map showing many linear features in the locations of the classical canals. The canals were also being discussed in reports of the British Astronomical Association at the good oppositions of 1939 and 1941—"Phison, seen occasionally. . . . Hiddekel, observed by several members. . . . Acampsis, rather narrow and faint" and so on. The question of the Martian canals still had not been resolved when Mariner 4, the first American spacecraft to be launched toward Mars, began its journey on November 28, 1964.

On July 14, 1965, Mariner 4 was 10,000 miles (16,090 km) from Mars. Automatically it started to take a series of 22 pictures over a period of 26 minutes as it flew past the planet. These pictures were stored in a tape recorder carried by the spacecraft. Eight and a half hours later, after the spacecraft had gone behind Mars as viewed from earth and then emerged on the other side, its automatic system began transmitting the stored pictures back to earth. Because of the limitations of radio receivers in those days, the data came back slowly, each picture requiring 8 hours to be received.

At the Jet Propulsion Laboratory (JPL) the data were processed by computer, and pictures of Mars were displayed on a TV screen. The first picture (figure 3.1) showed the bright limb, or edge, of the planet against the blackness of space. But no details were visible (much later the pictures were enhanced by computer processing to reveal some detail). Later pictures showed the surface only, but they too were very disappointing and gave no clues as to the nature of the Martian surface.

It was not until picture 5 was displayed that shadowy details could be seen. Many people were surprised to see a cratered landscape resembling that of the moon; they had expected Mars to be more like the earth. Although details were indistinct and diffuse as though out of focus, the pictures did show many craters, some of them large enough (150 miles across) to fill almost the whole of a photo frame. There was no sign of any of the types of Martian features so assiduously mapped by generations of astronomers. Scientists planning the mission had made a mistake in not scheduling some picture-taking to fill in between what can be seen from earth and what was seen close up; the Mariner 4 images could not be related to the Mars that astronomers saw through the telescope. Many scientists concluded that Mars was moonlike and was not, therefore, of great interest since the Apollo program would be scrutinizing the moon in great detail. The possibility of life on such a barren planet seemed remote.

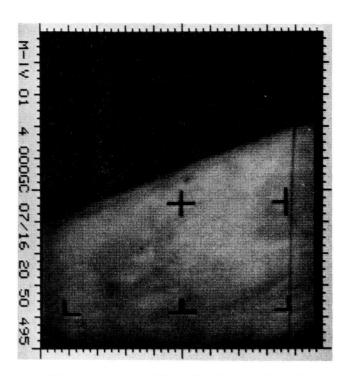

3.1 The first picture of Mars (the limb or edge of the planet) from a spacecraft was somewhat disappointing because it showed hardly any details. This version is a computer enhancement of the original picture, revealing faint shadings on the surface of the planet and some haze above the limb. The picture was taken at a distance of 10,500 miles (16,900 km) and shows the region of Phlegra. The spacecraft was Mariner 4, which flew by Mars in 1964.

The day after copies of the pictures were released, I inspected them closely for any sign that might provide a link between Mars as astronomers knew it and Mars as it appeared to Mariner. I applied several common photo-interpretation techniques. When the pictures were placed in correct relation to the best drawings of Mars made from earth, several of them straddled the boundary between the Mare Sirenum and the lighter area of Atlantis which lies between Mare Sirenum and Mare Cimmerium, at about 35 degrees south latitude. In this area the old maps showed a short linear feature, the canal Medusa. This corresponded to picture 11. Sure enough, on this picture I discovered linear markings in the gen-

The People and the Plan

eral direction expected of the dark streak of the canal (figure 3.2).

The markings looked to me like faults, possibly on either side of a rift valley; when I expressed this idea, however, the experimenters responded negatively. This discovery implied that the surface of Mars had not been molded solely by meteoric impacts but had also been subjected to forces from within the planet, perhaps similar to the tectonic forces that folded and faulted the crust of the earth. So I wrote letters to the editors of several science journals, which turned the

letter down because the official statement said that Mars was cratered and moonlike. After some correspondence, *Sky and Telescope* finally published my letter in January 1966.

Meanwhile local newspapers had picked up the story, and I wrote articles for the *Christian Science Monitor* and the British Interplanetary Society's *Spaceflight*. I also presented a short paper at the New York Academy of Sciences. But the crater syndrome was so strong that the idea of tectonic molding of the Martian crust was not generally acceptable.

Following the successful Mariner 4, the concept of an uninteresting moonlike Mars was so widespread that NASA did not fly any more spacecraft to Mars until 1969. Then Mariners 6 and 7 were sent to obtain more detailed information to complement details of the moon being returned by the Apollo program. The cameras of the new spacecraft were programmed to take a series of pictures that would link earth-based observations of Mars with detailed close-ups during a flyby. Each spacecraft carried a wide-angle and a telephoto camera. The wide-angle cameras started photographing when the spacecraft were millions of miles from the planet. The first pictures returned were similar to what astronomers see through their telescopes from earth. In subsequent weeks the pictures showed a larger and larger Mars and revealed ever more detail (figure 3.3). Then during the flyby the cameras produced swathes of pictures across the planet, showing a higher level of detail. Additionally, at many small areas within the swathes, the telephoto cameras produced images of very high resolution (figure 3.4).

3.2 The controversial frame number 11 obtained by Mariner 4 at an altitude of 7,800 miles (12,500 km) above Atlantis. This picture showed a lineament on Mars extending from the bottom left-hand corner to about one-third the way up the right side of the picture. A parallel lineament is just above it. The author was the first to point out that this might be a rift valley and evidence of tectonic activity on Mars.

None of these pictures showed anything that looked like a classical canal. Mariners 6 and 7 produced pictures that seemed to point to more and more craters and no tectonics. But the detailed pictures had covered only one-half of the planet.

The People and the Plan

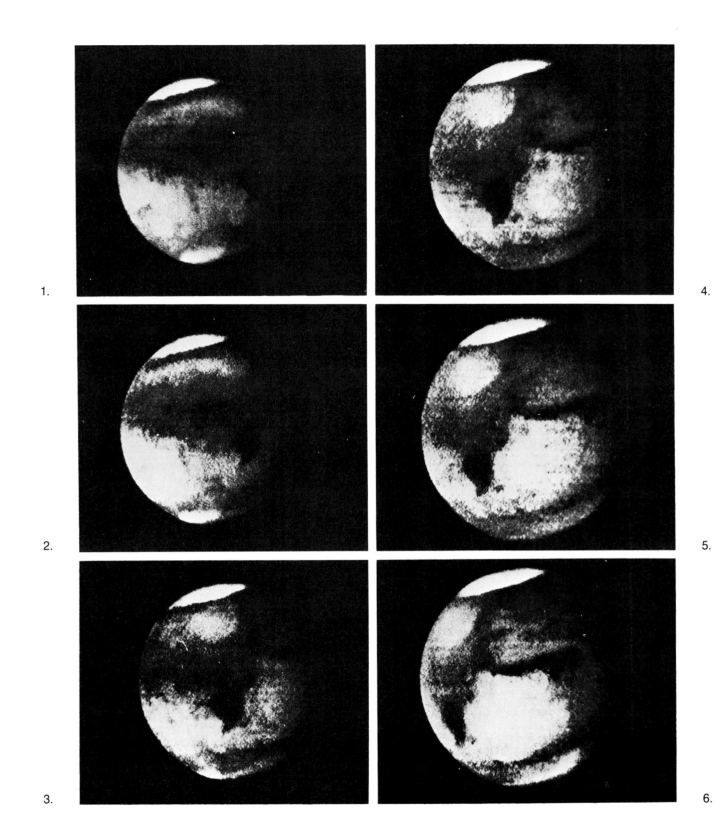

1.

2.

3.

4.

5.

6.

3.3 In 1969, Mariners 6 and 7 flew by Mars and obtained this series of approach pictures showing the revolution of the planet on its axis. The prominent dark triangular marking is Syrtis Major. South is at the top of these pictures.

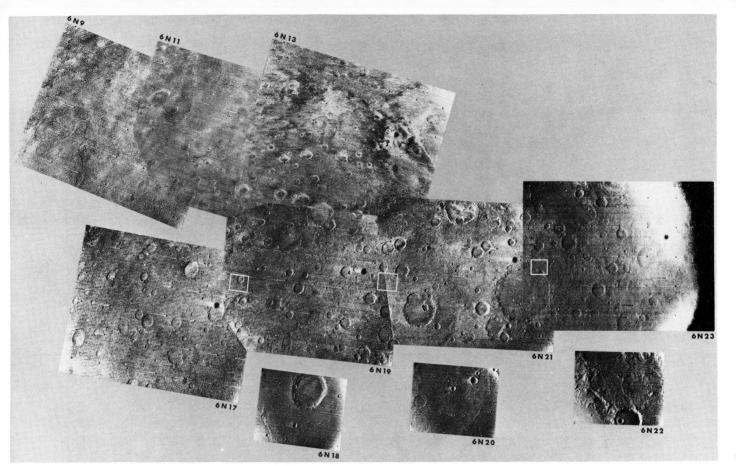

A.

B.

3.4 Mariners 6 and 7 also produced close-ups of the planet, together with high-resolution pictures as shown here. *A*, a general view across the Meridiani Sinus (0 degrees longitude); white boxes show the location of the high-resolution frames. *B*, an enlargement of the high-resolution frame 6N22 shown at the right bottom corner of the general view.

The question of whether the dark areas were vegetation was still unanswered. In 1956, William Sinton stimulated much speculation about life on Mars when he claimed that infrared spectra (beyond the red end of the visible spectrum) of the dark areas of Mars, obtained with the 200-inch (5-m) telescope at Mount Palomar, showed three absorption bands in a region of the spectrum where organic compounds with carbon–hydrogen bonds absorb. This observation was taken as strong evidence for Martian life. But by 1965, work by Donald Rea of the University of California showed that the absorption bands were caused not by organic molecules on Mars but by heavy water (formed of oxygen and the heavier deuterium isotope of hydrogen) in the earth's atmosphere. As a result the vegetation idea received a setback.

Generally, astronomers agreed that the bright areas of Mars were desertlike regions. Audouin Dollfus suggested in 1957 that on the basis of polarimetry and photometry, which measured the polarization and intensity of light from Mars, the bright areas were covered with limonite (a form of rust). Later, in 1964, R. A. Van Tassel and J. W. Salisbury suggested in a paper published in *Icarus* that the Martian deserts were more likely sand grains coated with limonite dust, rather than masses of limonite.

The nature of the dark areas produced much controversy. These areas cover about one-third of the Martian surface. Under high magnification and good telescopic visibility they resolve into groups of small dark spots rather than uniform areas. Observers reported seasonal and long-term color changes and variations in the extent and shape of the dark areas. Some claimed that a dark wave moved seasonally from each pole toward and 20 degrees across the equator. The most common explanation was that a wave of humidity passed from pole to equator as water sublimed (going immediately from a solid to a gas without melting into a liquid in between) from the polar cap in summer months.

Theories on the origin of changes to the dark areas fell into two groups—organic and inorganic. A popular inorganic, volcanic–wind theory, put forward by Dean McLaughlin in the 1950s, suggested that a planetwide wind system deposited ashes from active volcanoes in repeating patterns. Gerard Kuiper proposed that the dark areas might be lava beds from which lighter material was periodically swept by seasonal winds of Mars.

Organic theories attributed the color changes to growth of vegetation poking through dust deposits after the major dust storms. Speculations about the nature of this vegetation ranged from algae to higher plant forms.

Mariner 4 threw no light on the nature of the dark areas, nor did Mariners 6 and 7. The boundaries of these areas do not appear to correspond with any geological features visible in the pictures from the spacecraft.

In 1971 the second important event was that NASA readied two more spacecraft for flight to Mars. They were identical and were designed to orbit the Red Planet and fill in the missing details of areas not covered by the flybys of Mariners 6 and 7. The first of the two spacecraft, Mariner 8, was launched by an Atlas/Centaur booster on May 8. Lift-off was fine, and the second-stage Centaur separated from the Atlas and started to accelerate the spacecraft to the 24,600 miles per hour (36,000 km/hr) needed to reach the orbit of Mars.

Without warning the Centaur began to oscillate in pitch, and soon it was tumbling end over end. The Centaur dumped the Mars-bound spacecraft into the Atlantic Ocean. Engineers hurriedly analyzed data sent by radio from the launch rocket up to the time of failure, to find out its cause. There was little time if the second Mariner was to be launched during the 1971 launch window—the relatively short period when the positions of earth and Mars in their orbits would allow a spacecraft to travel from one to the other when launched by an Atlas/Centaur.

The engineers discovered that an integrated circuit was at fault, part of the electronics for the pitch-plane stabilizer of the Centaur. This tiny piece of electronics caused the loss of the complex spacecraft. Launch of the second Mariner was then scheduled for May 29, with arrival at Mars the following November. If the spacecraft could not be sent off by mid-June, the mission to Mars would be considerably delayed because the Atlas/Centaur was not a powerful enough booster to send Mariner to the Red Planet at the next opportunities in 1973 and 1975. In that case, no orbiting spacecraft would survey Mars before the planned Viking mission of 1975. This would have meant reconsidering the whole Viking expedition and perhaps changing it extensively, since there would have been no way to preselect landing sites before Viking arrived at Mars.

During this 1971 opportunity the Soviet Union also launched several spacecraft toward Mars. In earlier years, Soviet attempts to reach Mars had all been unsuccessful. Five spacecraft failed to leave orbit, and three missed Mars by wide margins. On August 2, 1965, Zond 2 reached Mars and passed within 1,000 miles (1,600 km) of the Red Planet, but it failed to return any data. At the 1971 opportunity a Soviet spacecraft launched on May 10 failed to separate from the upper stage of the booster rocket. Then Mars 2 was launched May 19; it orbited Mars successfully, returning data for about 3 months. On November 27, a capsule from Mars 2 landed on the Martian surface, the first spacecraft ever to do so. Unfortunately no data were returned after the landing.

Mars 3, which was launched May 28, 1971, reached the planet on December 2 and returned data from orbit for about 3 months. Its descent capsule landed

on Mars on December 2, south of Mare Sirenum, close to the border of Phaetontis, but only returned data for about 2 seconds. Both spacecraft had landed at the height of Martian dust storms, which may have caused their failures.

Meanwhile the second Mariner scheduled for the 1971 launch opportunity had encountered difficulties. Countdown was delayed because of problems in the ground support system, the complex electronic controls that ready the space vehicle for the launch. But on May 30 a successful launching took place at 3:32 P.M. Pacific Standard Time. I was in the Von Karman auditorium at the Jet Propulsion Laboratory for this launching, as were many engineers and their families from the Laboratory, where the spacecraft had been designed and built. The countdown was relayed to the auditorium from the Kennedy Space Center in Florida. Five minutes before the scheduled time to launch, the countdown was still going well. Then there was a hold because of a questionable readout on the propellant utilization system of the Atlas booster. "With all the layoffs in the aerospace industry, what can you expect?" muttered one of the engineers.

But it was a problem in the equipment on the ground and not within the booster, and the countdown soon resumed. At 3 minutes before launch the Centaur was declared to be in the launch mode. A short time later the big Atlas lifted majestically from its pad: all looked well. The Centaur engines ignited as scheduled, and this time the Centaur performed flawlessly. The engines cut off at an altitude of 110 miles (177 km), some 2,000 miles (3,200 km) down range. Mariner 9 separated and was at last on its way to the Red Planet.

The favorable Mars opposition took place during the flight of Mariner 9. The day after the opposition, on the evening of August 11, 1971, Mars drew closer to the earth than at any time since 1924. Its distance was reduced to 34.9 million miles (56.17 million km). Mars shone brighter than any star in the sky, brighter even than the planet Jupiter. It appeared as a reddish star in the constellation of Capricornus.

Just a few nights earlier I had stood in the darkened observatory where Percival Lowell made his historical and painstaking observations of Mars at the beginning of this century. Lowell Observatory is steeped in the lore of the planet Mars, and its vaults house a fine collection of Mars photographs and Mars literature. In preparation for the upcoming opposition and the arrival of Mariner 9 at Mars, I had spent a week at the observatory browsing through the records and talking with astronomers.

On this night Jay Inge, a planetary patrol observer (part of a team of astronomers located worldwide to photograph the planets on a continuing basis) was on duty at the 24-inch (61-cm) telescope that Lowell had had specially designed to study Mars. We watched the attached camera automatically click through a sequence of exposures as it photographed Mars through filters that passed different colors. Other similar cameras were organized in a network of telescopes spread around the earth to ensure that Mars was photographed almost continuously as it rotated on its axis.

When a faint whistle signaled the end of the exposure sequence, Inge motioned me to the eyepiece and I was able to look at Mars during the half hour before the next sequence of pictures was taken. I saw Mars just as Percival Lowell must have seen it at the opposition of 1894. The planet was a shimmering orange disc; the earth's atmosphere was not very steady because there had been thunderstorms during the evening. Mars appeared several times as big as a full moon in earth's sky. The south polar cap was prominent and brilliantly white, its edge surrounded by a dark collar; astronomers often reported seeing

this collar surrounding the cap as it shrank. Some dark rifts broke the smooth whiteness of the cap. Charles Capen, another experienced Mars observer who was with me at the time, explained that these dark areas appear year after year as the cap shrinks. On the south cap one prominent light patch becomes detached and is left isolated from the rest of the cap. It was called the Mountains of Mitchell, but astronomers were then generally undecided whether such white detached areas represent mountaintops or deep valleys.

These polar caps have been watched since the beginning of telescopic observation of Mars. They grow and shrink seasonally with great regularity. Each Martian year they begin to form in the northern or southern fall with the appearance of heavy white mists over the polar region. (On Mars, as on earth, the seasons are opposite in the northern and southern hemispheres.) Throughout the winter this hood of clouds covers the pole. As the winter ends, the hood begins to thin and the bright cap is revealed shining through it. Then the cap is seen at its maximum size. The south cap stretches 30 to 35 degrees northward from the pole; the northern cap is somewhat smaller, stretching only 25 to 30 degrees southward from its pole. From year to year, however, the maximum size of both caps varies.

During the summer a cap quickly shrinks in size; the southern cap often seems to disappear completely as seen from earth, although the northern cap does not. The reason is that summer in the Martian southern hemisphere occurs when the planet is close to perihelion, whereas the north polar region receives its summer heating when Mars is farther from the sun and close to aphelion. Although the north cap is almost centered about the north geographic pole of Mars, the southern cap is displaced about 6 degrees, probably because of local topography.

The People and the Plan

The rapid disappearance of the caps suggested that they were extremely thin. Some estimates were that the thickness could not be more than 0.4 inch (1 cm) on the average. But the central part that does not disappear could be much thicker. Originally many astronomers thought the polar caps consisted of water ice, analogous to earth's frozen polar wastes. Others thought they might be frozen carbon dioxide. In 1948 Gerard Kuiper compared their infrared spectra with laboratory samples and concluded that the caps consisted of water ice.

A major question about Mars was how cold it might be, because this would show whether its polar caps were carbon dioxide or water ice. Polar temperatures of −170 degrees Fahrenheit (−112°C) or warmer would indicate that the polar caps were water ice. Scientists were hoping that the infrared radiometer experiments of Mariners 6 and 7 would answer the question. The infrared radiometer was designed to measure the temperature of the surface of Mars from its infrared (heat) emission. But the experiments did not really resolve the matter. Infrared scans across the south polar cap indicated a temperature there of −189 degrees Fahrenheit (−123°C). This pointed strongly to the cap being solid carbon dioxide, not water ice. Measurements by another instrument indicated, however, that surface temperatures were too high for a solid carbon dioxide cap. Although the experiment recorded solid carbon dioxide in its scan across the cap, this was interpreted as suspended carbon dioxide ice particles in the atmosphere, not on the surface.

The material of the polar caps was important because if the caps were solid carbon dioxide they could store a reservoir of this gas. A general warming of the planet could release the gas and considerably increase the pressure of the Martian atmosphere. This would be important to the question of life on Mars, since a greater atmospheric pressure would mean that water could flow periodically as a liquid on the Martian surface.

Estimates of the proportion of carbon dioxide in the atmosphere of Mars varied greatly over the years. In 1949, Dr. Kuiper calculated that there was twice as much carbon dioxide in the atmosphere of Mars as in the atmosphere of the earth. Then in 1955 J. Grandjean and R. Goody calculated that it was 13 times as much. They based their estimates on a total surface pressure of 85 millibars (less than one-tenth that of the earth's atmosphere at sea level). The surface pressure obtained by Audouin Dollfus in 1957 from interpretation of polarimetry was between 80 and 90 millibars. Even in 1961 the surface pressure of Mars was still assumed to be 85 millibars.

By 1964, however, as a result of improved spectroscopes, the Martian atmospheric pressure was accepted as being only 17 millibars. A combination of a specially sensitized film and an improved spectrograph at the Mount Wilson 100-inch (2.54-m) telescope yielded an unusually clear spectrogram of Mars. A high-resolution spectrograph divides light into its various frequencies and records light intensity at each frequency as a series of lines that form bands. Inspection of the spectrogram revealed the presence of a band of carbon dioxide so weak that it had never before been detected. It allowed a direct measurement of carbon dioxide in the Martian atmosphere that was incompatible with previously accepted estimates of the Martian atmospheric pressure. This measurement was made by Lewis Kaplan, Guido Munch, and Hyron Spinrad, combining their results with measurements of carbon dioxide and atmospheric pressure made by William Sinton in other regions of the spectrum. The constituents were thought to be 85 percent nitrogen, 14 percent carbon dioxide, and 1.0 percent argon, with mere traces of water and oxygen.

The People and the Plan

When Mariner 4 passed behind Mars during 1964, the spacecraft's radio signals to Earth were temporarily interrupted by the bulk of the planet. This occultation provided information about the atmosphere of Mars; namely, that it contained more carbon dioxide than nitrogen. The Martian atmosphere was now estimated to be 80 percent carbon dioxide and the rest nitrogen and argon. More important still to the possibility of life on Mars, the flyby of Mariner 4 suggested that the surface pressure was only 4 to 7 millibars. This is equivalent to the pressure 21 miles (34 km) up in the earth's atmosphere.

Persistent efforts from 1900 onward had failed to detect molecular oxygen in the atmosphere of Mars. Water vapor, however, was regarded as a constituent from the time the polar caps were first seen. Despite earlier claims to the contrary, water vapor was not detected from the earth, even at the good opposition of 1956. In 1963 Kaplan, Munch, and Spinrad, again using the 100-inch (2.54-m) telescope at Mount Wilson Observatory, obtained a high-resolution spectrum of Mars in which the absorption lines were more clearly separated from one another. They were able to estimate that a very small quantity of water vapor might account for absorption features that they saw in the spectrum. They calculated that there was 1,000 to 2,000 times less water vapor in the Martian atmosphere than in the earth's. If condensed on the surface of the planet (with a diameter of 4,260 miles, 6,856 km) the water would form a layer covering it to a depth of less than a thousandth of an inch (0.03 mm). Equipment carried to a high altitude by a balloon to reduce the effects of water vapor in the earth's atmosphere confirmed the results.

The really conclusive proof of water on Mars came in March 1969, when astronomers at McDonald Observatory at the University of Texas obtained spectra with the 82-inch (208-cm) Struve reflecting telescope and its large spectrograph. There were several extremely dry days at the desert location of the observatory, so that the small amounts of water vapor in the earth's atmosphere did not mask the Martian water vapor. Also, Mars was moving rapidly toward the earth, and the relative motions of the two planets shifted the spectral lines due to water vapor on Mars in relation to those resulting from water vapor in our atmosphere and made it easier to identify them.

Water vapor was measured as being equivalent to a film of water about 2 thousandths of an inch thick (0.05 mm or 51 precipitable microns, the depth of a layer of liquid water that would result if all the water vapor in the atmosphere condensed) in the southern hemisphere of Mars. This corresponded to a little over 1 cubic mile (4.17 cu. km) of water in the whole atmosphere of Mars. (By comparison, air over the Arizona desert contains about 1 millimeter of precipitable water—20 times the amount of water that had been found in the Martian atmosphere.) These results implied that the caps must contain some water ice as well as carbon dioxide ice.

At the end of the 1971 opposition, Mars did not look very hospitable to life. Dr. Norman Horowitz had commented soon after the scientific results from Mariners 6 and 7 had been assessed: "There is nothing in the data returned by Mariner to show that there is life on Mars. It is a desolate planet with a thin dry atmosphere consisting largely of carbon dioxide, and unfiltered solar ultraviolet light penetrates to the surface. No terrestrial species could live there."

Meanwhile, as Mariner 9 traveled toward its rendezvous, Mars had started to develop an enormous storm that enveloped the planet (figure 3.5). I asked William Baum, director of the planetary patrol program at the Lowell Observatory, about the storm in view of Mariner 9's approach to Mars. Dr. Baum said that a yellow cloud, presumably of dust, had appeared on

The People and the Plan

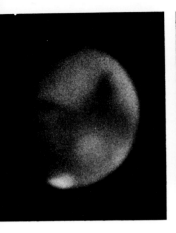 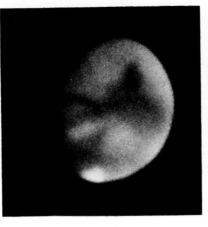 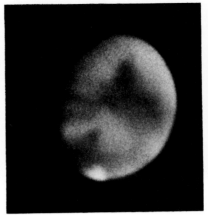 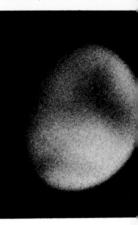

Sept. 21   **22**   **23**   **Oct. 3**

3.5 Development of the planetwide dust storm following the 1971 opposition. This series of pictures, taken in green light, shows the storm beginning west of Hellas on September 22, and spreading to engulf most of the southern hemisphere of the planet by October 3. The triangular-shaped dark marking is Syrtis Major, and the planet is oriented with the south pole at the bottom.

September 22 in the dark region of Mare Serpentis and the light-colored Noachis basin, which are west of the large circular depression known as Hellas. "This type of cloud is expected each Martian year about this time," he explained. At first the cloud (which obscured all surface detail) covered only a small part of Mars. After a few days, it suddenly expanded and cloaked the whole southern part of the disc of Mars visible from earth. The bright southern polar cap was also enshrouded. Clouds of this type, explained Dr. Baum, are common when Mars is closest to the sun. They have been recorded by astronomers since 1877. Such dust clouds usually seem to start from the same region of Mars, just west of Hellas, and abruptly envelop the whole planet a short while later.

At the 1956 perihelic opposition, the clouds prevented astronomers from seeing any details on the surface while Mars was at its closest to the earth. In the 1971 opposition, however, the clouds did not cover the planet until after perihelion and after opposition.

The storm was still blanketing Mars when Mariner 9 arrived and was placed in orbit on November 13, 1971. The first pictures returned by the spacecraft looked like photographs of an old tennis ball. The storm obscured Mars for several weeks after Mariner 9's arrival. As the dust gradually settled out of the Martian atmosphere onto the surface, first the polar caps and then four dark spots appeared.

The dark spots proved to be the calderas (collapsed basinlike central craters) of four huge volcanoes in the area of Tharsis—volcanoes that were quite unexpected from earlier flybys of Mars. Nix Olympica (snows of Olympus), a light area on maps of Mars, turned out to be the largest volcano known. Its summit towers some 17 miles (27.4 km) above the surrounding plain, and its base measures 310 miles (500 km) across (figure 3.6). It was renamed Olympus Mons (Mount Olympus) when its true nature was recognized. Mariner 9 also discovered a system of great canyons on Mars, into whose tributaries the earth's Grand Canyon would fit with room to spare (figure 3.7).

Mariner 9 revealed the extensive nature of the crustal faults I discovered on Mariner 4's picture 11. The rift valley through the large crater of Mariner 4's picture was seen again and was found to be about 2,300 miles (3,700 km) long. It was named Sirenum Fossae (*fossa* is Latin for a riverbed or watercourse).

The People and the Plan   33

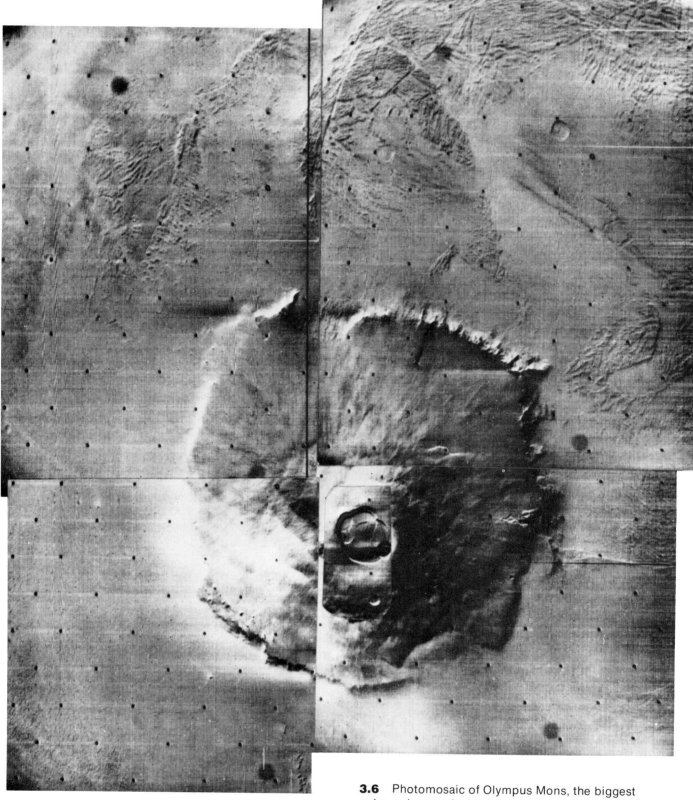

3.6 Photomosaic of Olympus Mons, the biggest volcano known. Its base is 310 miles (500 km) across, and its summit is 17 miles (27.4 km) above the surrounding great plain. Steep cliffs drop off from the mountain flanks to the plain, on which there are features very similar to those on the floors of terrestrial oceans. The caldera on the summit of this shield volcano is 43 miles (70 km) across.

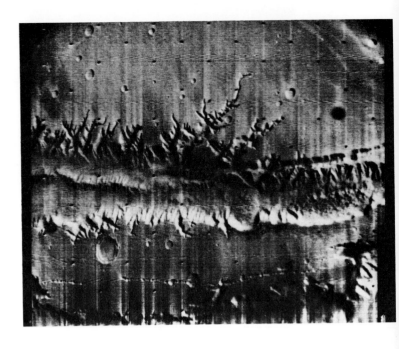

3.7 Mariner 9 revealed a vast canyon system on Mars, of which only a part is shown in this picture. This section has a spinelike ridge running along its floor, and branching tributaries as large as the Grand Canyon of Arizona.

But Sirenum Fossae does not correspond to any classical canal; on the drawings by astronomers who claimed to see canals, Medusa actually runs in a somewhat different direction. My discovery of Sirenum Fossae had been fortuitous.

In the following months of observations from orbit, Mariner 9 determined that Mars is pear-shaped, with its southern hemisphere bulging outward and its northern hemisphere pushed inward relative to a perfect sphere. The lowest regions of Mars were in a band at about 65 degrees north latitude. The north polar region itself was close to mean level. This discovery of low-lying areas in northern latitudes was important to the selection of landing sites for Viking, because low-lying areas might possess sufficient atmospheric pressure for water to be a liquid at the Martian surface.

The southern highland regions of Mars were heavily cratered; probably the cratering dates from shortly after the formation of the planet. The lowland regions, by contrast, contained relatively few craters and appeared to be covered by extensive lava flows and other deposits of various ages. The boundary between high and low regions approximated a circle around the planet but did not coincide with the equator. It was tilted some 35 degrees from Mars' equatorial plane.

Each polar region seemed to be covered by layered deposits possibly resulting from different epochs in the history of the planet. Material eroded from the desposits by wind appeared to have migrated toward the equator, producing blankets of materials covering the temperate regions of the planet. There may also have been extensive erosion of equatorial regions to provide the material that was deposited at the poles, possibly entrapped in snowfalls there.

Mariner 9's major discovery was the many channels on the Martian surface—broad channels, sinuous channels with tributaries, and textured channel networks (figure 3.8). There also appeared to be flood plains across which major flows of liquid had taken place and caused intertwining broad channels and streamlined islands. The only reasonable explanation for the channels seemed to be that water once flowed in large amounts on the surface of Mars. Some channels, however, had the characteristics of terrestrial and lunar lava channels.

Photographs showed wind erosion on a broad scale over the whole of the Martian surface. Also they suggested that a process of differentiation had taken place on Mars similar to that on the earth, with formation of a crust and possibly a core deep within the planet. But major questions remained to be answered, such as the chemical composition of Mars. The varied surface features seen from Mariner 9 suggested that Mars had a surface crust of variable thickness and that the rocks on the surface of this crust were possibly of different substances and different ages.

The People and the Plan

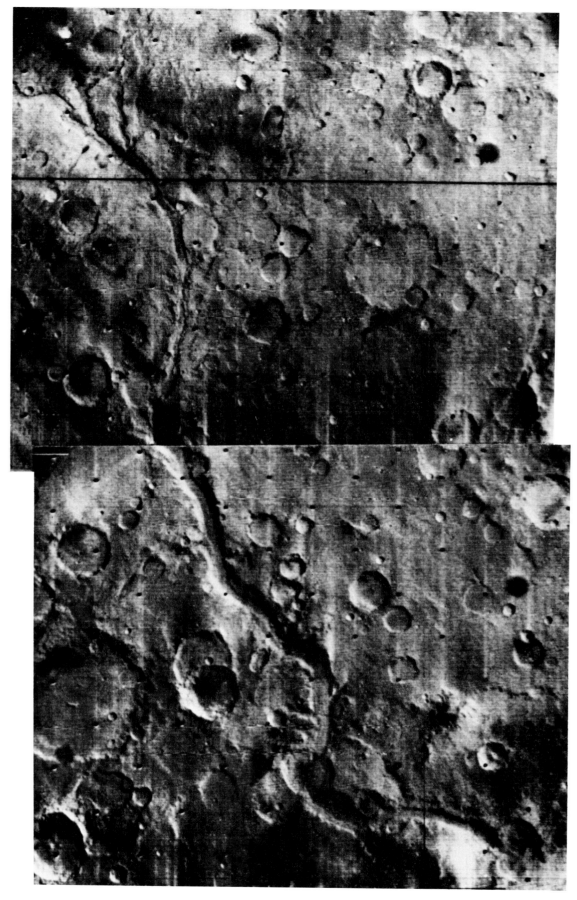

3.8 A major discovery of Mariner 9 was channels on Mars that looked very much as though they had been produced by flowing water. There are multiple branched tributaries at the head (*at left*), and the channel deepens toward its mouth. This channel is located in heavily cratered terrain south of Apollinaris Patera, and is named Ma'adim Vallis.

There was speculation that changes to the orbit of Mars and wobble of the polar axis could melt carbon dioxide reservoirs in the polar caps and periodically increase the Martian atmospheric pressure sufficiently for water to flow and for rains to fall. Such speculation could only be resolved by a mission to land a spacecraft on Mars, sample the surface, and obtain better information in general about the Red Planet. This was to be Viking.

In addition to the excellent opposition and the launching of Mariner 9, 1971 was important because the Viking project became fully active again after two years of reduced funding. Originally scheduled for a 1973 launching toward Mars, Viking had by this time been replanned, because of budget limitations, for a 1975 launch so that the two spacecraft would arrive at the Red Planet in 1976, the Bicentennial.

Viking actually began as a concept in the mid-1960s when plans were taking shape for missions to land capsules on other planets. Originally the landing mission was conceived as using the Saturn V booster that was being developed for the moon landings. With such a booster, fairly large payloads could be landed on Venus and Mars, and sent to orbit other planets. The initial concept for the spacecraft was called Voyager, and the program was regarded by NASA officials as the most important undertaking in space exploration since Apollo began in 1961.

The program ran into difficulties arising from the aerospace cutbacks and a so-called reordering of national priorities in the late 1960s to adapt to the economic problems associated with the disastrous war in Southeast Asia. In 1967, although considerable sums had already been spent on Voyager, aiming for a mission to Mars in 1973, Congress decided that there would be no major increases in planetary programs during the year; Voyager had effectively been cancelled.

NASA next concentrated on finding a way to land a sufficient payload on Mars to make a life-seeking expedition worthwhile, but on a much smaller budget. The decision was made to use a military booster (the Titan III) with a Centaur upper stage as the launch vehicle, instead of Saturn V. This would reduce the cost of the mission considerably, but it meant that a smaller spacecraft would have to do the job. This spacecraft was conceived as an orbiter, derived from the Mariner series of spacecraft, that would carry a lander derived from Surveyor, the first spacecraft to touch down on the surface of the moon. The lander would soft-land on Mars and would sample and photograph the surface. It would also carry an instrument to seek evidence of life.

Initial studies showed that the mission was feasible and very worthwhile. As a consequence, President Lyndon Johnson announced the new mission in his 1968 budget message to Congress. Later that year NASA released details of the project to place two spacecraft on the Martian surface in 1973.

There would be two launches, each of an orbiter carrying a lander—that is, four spacecraft in all. Each orbiter would be placed into orbit around Mars; when a landing site had been certified as safe, it would release its lander, which would then descend to the surface. Speeding through the Martian atmosphere, the lander would be slowed down by atmospheric braking and then use terminal-descent rocket engines for a touchdown. Each lander would be designed to operate on the surface of Mars for at least 90 days. The orbiters would also remain in orbit for at least that period to gather data about the planet and to act as communications relays from the landers to earth.

Viking was thus the most complex space mission to be planned by NASA. It required the simultaneous operation of four spacecraft an an enormous distance

The People and the Plan

of nearly 200 million miles (320 million km), because Mars would be on the far side of its orbit from the earth when the four spacecraft reached it.

NASA selected the Martin Marietta Corporation, in Denver, to design and fabricate the lander spacecraft and to provide technical integration of the mission. The Jet Propulsion Laboratory, in Pasadena, was selected to build the orbiter spacecraft and to provide tracking and communications support for the mission, together with facilities for controlling the operations of the mission. Overall management of the Viking project was allocated to NASA's Langley Research Center, in Hampton, Virginia. The project manager was to be James Martin, a space-mission veteran who had managed the trail-blazing mapping of the moon by several unmanned spacecraft. These lunar orbiters photographed the surface of the moon in great detail to provide information needed for the landings of the Apollo astronauts.

NASA also appointed a science steering group, so that scientists could participate early in the design of the spacecraft and of the mission itself. Later, science teams were formed for each experiment, and they became closely involved with all developments of the scientific instruments to be carried by the spacecraft to Mars.

Viking was not solely an expedition to search for life on Mars, but the search for extraterrestrial life was undoubtedly a major factor in obtaining public and congressional support for the project. The possibility of searching for life elsewhere in the solar system had been stimulated for a decade or so before Viking by a handful of scientists, some of whom actually became experimenters on the Viking mission.

Among the scientists who first became involved in the design of experiments to search for extraterrestrial life was Joshua Lederberg of Stanford Medical Center, who inspired many other people to become interested in and devote their time to the study of exobiology. In 1959 he had received a small contract from NASA to conduct a 1-year study of cell chemistry pertaining to possible extraterrestrial organisms. Two years later Dr. Lederberg defined the objective of exobiology as the comparison of patterns of chemical evolution on the different planets. He said that the question of life on Mars dominated all speculation about exobiology. He favored the search for microorganisms at the surface rather than observations from orbit. To look for these microbes he suggested that a remote microscope, telemetering its images back to earth by radio, was the most efficient sensory instrument. However, other exobiologists were quick to point out that pictures can be misleading. Specimens of rocks and sediments on earth contain microspheres, inorganic filaments, and globules that are not living systems. From pictures alone it is impossible to decide whether these forms are microbes.

Working with Dr. Lederberg at Stanford was Elliott Levinthal. He, too, had been interested for some time in the search for extraterrestrial life. In 1962 Dr. Levinthal became principal investigator of a biomedical instrumentation laboratory at Stanford with the aim of designing devices that might detect life on other planets of the solar system. Addressing a seminar at TRW Space Systems in Redondo Beach, California, in November 1963, Dr. Levinthal spoke about means to detect life on Mars: "If something moves in a direction different from the wind, or under a microscope moves in a non-Brownian [nonrandom] fashion in a direction other than the fluid flow, it would be expected that this something was alive or propelled as the result of some activity of a living system. That would be called a sign of life, and instruments could be designed to use that criterion on a macro or micro scale." But he added that observations by such instruments would not be of much use in determining the type of life, and how it differed from life on earth.

The People and the Plan

Dr. Levinthal went on to say that the most prevalent form of life was likely to be microbial. He discussed the use of biochemical amplifiers to detect such life; a few enzyme molecules are detectable because they can catalyze reactions of other molecules at the rate of tens of thousands per minute. If a bacterium grows and divides every 30 minutes, it will produce in 10 hours, under the correct conditions, more than 1,000,000 copies of itself. This proliferation, if detected, could also be used as a sign of life.

Early in the 1960s, Levinthal and Lederberg devised one of Stanford's first experiments in exobiology using the principle of detecting an enzyme by the fluorescence of its residue. They called their instrument a multivator. Later they were more concerned with developing concepts that other scientists could convert into actual hardware. Lederberg warned that no experiment could be certain to produce results quickly. It might have to operate on Mars for weeks, possibly months, before a positive reaction could be obtained. Lederberg also said that a camera might be the most important of the experiments searching for evidence of living organisms, past or present, on the Red Planet.

Another microbiologist who became excited about exobiology at about that time was Wolf Vishniac, of Brookhaven National Laboratory on Long Island. Dr. Vishniac believed that the search for Martian life should not be restricted to looking for any specific organism. Wherever there exists an environment that is populated by living organisms, it must have more than one type of organism, he claimed. There must be a combination of electron donors and electron acceptors—the equivalent of the animals and plants of our own earth.

In June 1959, Dr. Vishniac received a small, 1-year contract from NASA to develop an experiment for remote detection of bacteria by automatic observation

of changes in a culture medium, using what he termed a "Wolf trap." The instrument was designed to suck dust from the surface of Mars and put it through an experiment to reveal any earth-type organisms in the dust.

In a paper published in 1972 in the scientific journal *Icarus,* Dr. Vishniac described this life-detection apparatus. He stated his belief that the most conclusive evidence for the presence of life on a planet such as Mars would be a simultaneous observation of increased metabolic activity and growth. Regrettably Dr. Vishniac's life-detection equipment was canceled for Viking during a period when costs had to be cut to meet the reduced NASA budget. Later this pioneer of exobiology lost his life in an accident while exploring the valleys of Antarctica to determine how terrestrial microbes have adapted to adverse environmental conditions.

Another biologist, Gilbert Levin, worked independently on ways to detect life on Mars by observing the effects of microbial growth and metabolism. In 1965 Levin, then with Hazleton Laboratories in Falls Church, Virginia, spoke at a meeting of the American Astronautical Society. He observed that "if life is not found we shall want to know why it has not arisen and whether the process of prebiological evolution is operating. The information obtained may materially aid our theories concerning the origin of life on earth."

By 1967 Levin had designed a microbial metabolism laboratory that was automated to make experiments on Mars. Levin said that he preferred a "ferret approach" to the search for extraterrestrial life, rather than remote-sensing approaches. He wanted to drag Martian life from its hiding places and inspect it with an apparatus that tempted the microorganisms with tidbits of "food" labeled with radioactive tracers. In his experiment, named "Gulliver," he assumed that life on Mars would be carbon based and that its biochemical reactions would take place in water. He picked the name Gulliver because Swift's hero was a person who traveled to far places searching for strange beings. Later, after the Viking landing on Mars, Dr. Levin commented: "I never realized how strange some of these beings might be or how difficult it might be to try and find them."

He took one of his instruments to a seemingly barren slope on White Mountain, in California, and in the midst of a snowstorm it detected the presence of microorganisms within less than an hour. This performance was repeated on the desolate sand dunes of Death Valley and on the salt flats adjacent to California's Salton Sea.

I first met geneticist Norman Horowitz of the California Institute of Technology in September 1959, when he chaired a session on astrobiology for a group to which I belonged. The Lunar and Planetary Exploration Colloquium held meetings in the Los Angeles area, mainly sponsored by the aerospace companies, to discuss in practical terms how the moon and the planets might be explored. The members included astronomers, geologists, biologists, and technical managers like myself, as well as engineers and other scientists connected with the aerospace business. At the symposium on astrobiology Dr. Horowitz asked the members to discuss methods that might be used to detect life on another planet. Some biologists expressed concern about the type of nutrient medium that would be acceptable to a Martian organism. Dr. Horowitz pointed out that the hypothesis of a universal DNA–protein form of life provided an answer. "If Mars has a DNA–protein form of life, we would expect components of protein and DNA to be essential nutrients."

I did not meet Dr. Horowitz again until 1973. He was then one of the official experimenters for Viking. His life-seeking experiment, directed toward looking for photosynthesizing organisms, had been accepted as part of the biology package to be flown to Mars. I talked with him about possibilities of life on other worlds. What we are trying to discover, he said, is the history of life in the solar system. By exploring Mars we can learn more about the origin of life than we can learn in any other way, even if Mars is uninhabited. The program will by no means be unsuccessful if we find there is no life on Mars. Viking is a voyage of discovery and of exploration.

There were, of course, a number of life-seeking instruments on Viking; the TV camera, for example, could reveal the presence of any plant life. Atmospheric chemistry is important to life, and other experiments could determine what gases there were in the atmosphere of Mars. Another instrument was to analyze the soil, looking for organic substances, and there were several direct biological experiments among which Dr. Horowitz's was designed to look for incorporation of carbon (from atmospheric carbon dioxide or carbon monoxide) into a living organism. The experiment operated by supplying carbon dioxide and carbon monoxide labeled with radioactive carbon 14 and checking whether any of it was picked up by a Martian microorganism, either when exposed to simulated Martian sunlight or in the dark.

Asked about fossil life, Dr. Horowitz said that although in principle the TV camera could detect artifacts such as petrified vegetation, the organic analysis of the soil was important because it should reveal if any complex organic chemistry had existed on Mars in the past. Also, he said, it was very important to find out if anything now lives on Mars.

Another biologist involved in Viking was Vance Oyama, whom I met in 1974 while working on a book for NASA on the Pioneer mission to Jupiter. Dr. Oyama had conducted a whole series of experiments

The People and the Plan

on lunar soil samples, but had sought in vain for evidence of present or past life on the moon. Later he developed a life-seeking experiment for Viking. Dr. Oyama's experiment was essentially seeking evidence in Martian soil of the ability of living things to exchange gases with their surroundings on a continuing basis, as opposed to the depletion of chemical reactants in a soil sample.

In Dr. Oyama's experiment, water was to be provided along with organic materials to make a nutrient solution. The exchange of gases in the air space of a test chamber, into which Martian soil would be placed and wetted with the nutrient solution, would be checked over several weeks of incubation. Dr. Oyama said that he had to decide what could be considered a biological response and what would have to be accepted as a nonliving chemical process. So he looked at as many minerals and elements as he could to determine the kinds of chemical changes that occur during a typical incubation period. He also checked the responses of many soils containing microorganisms.

The criterion for finding life was agreed upon as finding growth. That is, if a nutrient was fed into the soil and the nutrient continued to give rise to a gas exchange by the soil sample, the presence of a growing, living organism could be accepted. But if, after a while, the addition of more nutrient produced smaller and smaller gas exchanges, this would have to be taken as evidence that some chemical reaction was occurring in the soil, and that the chemical responsible for gas exchange was being used up by the addition of nutrient.

The three life-seeking experiments of Levin, Horowitz, and Oyama were integrated for Viking into a combined biological package. This task was performed by TRW Systems, managed by NASA's Ames Research Center, where Dr. Harold Klein had responsibility for the biology experiments for the Viking expedition.

I met "Chuck" Klein in June 1971 at a meeting of COSPAR in Seattle. COSPAR is an acronym for the Committee on Space Research of the International Council of Scientific Unions. It was organized in October 1958 to help international collaboration in space sciences, and it meets annually in different countries. The American Astronautical Society held one of its West Coast meetings in Seattle to coincide with the COSPAR meeting, and Dr. Klein gave a talk, coauthored with Wolf Vishniac, on biological instrumentation for the Viking mission. In an interview afterward Dr. Klein said that the combination of four active biology experiments (Dr. Vishniac's experiment was then still planned for Viking) covered a wide range of assumptions about a possible Martian biota (plants and animals), from dry to wet experiments and from autotrophic to heterotrophic conditions. Dr. Kein continued: "From a biological point of view the Viking '75 mission can be regarded primarily as a test of the Oparin–Haldane hypothesis of chemical evolution. Findings of amino acids and hydrocarbons in the Murchison and Murray meteorites, as well as the recent radioastronomical findings of formaldehyde, hydrogen cyanide, methanol, cyanoacetylene, and other organic compounds in the interstellar medium, all suggest very active chemical processes leading to the formation of compounds of great biological and chemical interest." He then added, "If evidence is obtained suggesting a Martian biota, this will clearly lead to new insights concerning current theories on the origin of life, and on the fundamental properties of living systems."

The next time I spoke with Dr. Klein was in 1975, at about the time when the biological package was in final tests and the deadline was approaching for two units to be incorporated into the waiting Viking landers. Asked about his thoughts concerning the

The People and the Plan

possibilities of life on Mars, he responded that the members of the biological team had a wide divergence of opinion. He himself thought the chance was low—"I believe I've come up with the number of one chance in fifty."

He said that although he was pessimistic about there being life on Mars, he still was open-minded about it, "because we've been surprised about Mars by Mariner 9 data. I look for great surprises. We may see something that of itself may be even more important than the chemical or biological test. None of us knows what is really on Mars. But what is more important is that even we who are on the pessimistic side are absolutely convinced that the payoff if we are wrong and there is life, is enormous."

Why was Dr. Klein pessimistic about the possibilities of Viking finding life on Mars?

> For this argument I come down to two things. First, the whole thing hinges on a faith in the theory of chemical evolution. This has a lot of things going for it, but it is by no means proved. If it were not for that [theory], we would just be clutching at mysterious reasons to say that there should be life on Mars.
>
> Second, we go from that to ask how long did it take for life to develop on earth, assuming that the theory is right? We come up with estimates that it took very little time; in the first half billion years.
>
> If Mars and earth were spawned at the same time, of the same primordial material, the two planets would have been more similar then than they are today. There is a very good chance that it [life] started on Mars at the same time and in about the same way.
>
> Then, if it ever got to a replicating system when Mars had a much denser atmosphere, when the two planets diverged over the next two or three billion years, organisms would have evolved as they went along.

But Dr. Klein said that when he considered Mars as it is today, and what is known about the limitations of terrestrial life, he could not really visualize a sustained biota on a planet that is so dry and is bathed in fierce solar ultraviolet radiation. He added that if he were wrong and Viking did find life on Mars, "the scientific and philosophical payoff is so great that it is like a two dollar bet on the Irish Sweepstakes."

Early in the 1970s I had been writing series of educational materials (The Now Frontier) for NASA for use by teachers in high schools, and these had been received very well. The idea had originated in the public information department at the Ames Research Center, for which I prepared material on the Pioneer mission to Jupiter.

I received a contract to write a similar series on Mars and Viking (which appeared as NASA Facts), and in the fall of 1974 I traveled to the Martin Marietta facility in Denver to see how the construction of the spacecraft was proceeding. It had been over 15 years since my first visit to the Martin plant. That had been at the height of the ballistic missile program, when I was working on a book about long-range missiles. Then the test stands in the hills behind the plant had been actively testing mighty rocket engines for the Titan ICBM. I was a little sad to see rusting metal and cracked concrete chutes in place of the intense activity of those earlier years.

The action was now elsewhere. Big rocket engines were commonplace and no longer needed test programs to perfect them. The old stands had largely been recycled as scrap metal. Now the engine development of the late 1950s was making it possible for a highly sophisticated machine, loaded with computers and instruments, to be sent to land on a distant planet. And, indeed, it would be a Titan, though a much updated Titan III, that would launch the Viking spacecraft on its long voyage to Mars.

On October 28, Veteran's Day, I donned a white "bunny suit" and overshoes to enter the "clean

The People and the Plan

3.9 A complex machine to land on Mars, one of two Viking Lander spacecraft, is here shown at the assembly facility of Martin Marietta Aerospace, Denver, Colorado. All workers and visitors wore protective clothing to reduce contamination of the spacecraft and its parts.

(PHOTO, MARTIN MARIETTA AEROSPACE)

rooms" to visit the almost completed spacecraft under its final tests. It was a fantastic experience to be standing in that huge, hangar like room among white-coated men and women who were bustling around an ungainly object on three short legs—a spacecraft designed to land on Mars (figure 3.9). Cables snaked around the Viking lander, and everything gleamed under the harsh light of the fluorescents high overhead. I could reach out and touch a machine that was actually going to land on Mars.

From the Martin plant I traveled to the Langley Research Center in Virgina, where I met Dr. Gerald Soffen. Gerry Soffen impressed me immediately. I saw him as a person of tremendous dedication and unbounded enthusiasm for this expedition to Mars. I imagine that if I had been able to talk to Columbus, or Lewis and Clarke, or Livingstone, I might have had the same impression. Here was a relatively young man who had devoted all his professional career to the question of whether other worlds of the solar system had developed living things like the earth. His dedication and technical skills had won him what might be the prime scientific task of this generation—acting as chief scientist of a project to land the nation's first spacecraft on Mars, in fact, the first American spacecraft to land on any other planet. (The moon, of course, is a satellite of earth and not a separate planet, although because of the moon's size this is sometimes regarded as a two-planet system.)

In his quiet manner, Dr. Soffen recounted how it all came about: "I can remember one time in New York City there was a meeting of the New York Academy of Sciences. I was a young graduate student, and the professor I was working under said I might be interested in what Harold Urey and one of his graduate students were doing in trying to make amino acids. It was 1956. I went to listen to Stanley Miller, and was overwhelmed by the idea, he could actually make amino acids from simple gases! It sounded almost

unreal. It seemed like a crack in the system [of then-current biology], . . . one might really get a glimpse of how the chemistry [of life] is done."

Dr. Soffen said it was by chance that the prebiological chemistry happened to be demonstrated at the right time to stimulate the whole biological impetus of the space program.

In 1961, continued Dr. Soffen, he was working on enzymes at New York University, and one evening, because he was curious, he walked into a meeting of the Institute of Radio Engineers. Asked by a young man there if he was an engineer, he said no.

But before I knew it, I was being interviewed by two engineers who asked me if I was interested in life on Mars. This was the first time I had heard anyone use such an expression. I replied, "No, what are you talking about?" They then asked me if I was an exobiologist, and I said I'd never heard the term.

They told me it was a term that Josh Lederberg was using for extraterrestrial biology. And out of curiosity—only curiosity, for at that time I was working on biochemistry and expected to do that for the rest of my life—I found myself across the country being interviewed at the Jet Propulsion Laboratory and talking about life on Mars. The whole thing seemed absolutely unreal; it was 1961, and these people were talking about sending a rocket to Mars! I thanked them; but one thing stuck in my mind. They had mentioned that Norm Horowitz was interested in it. So I went to Caltech before I came home, and I

The People and the Plan

talked with Horowitz. He was really serious. It was the first time I had really met a visionary in biology, a man who saw the future rather than the immediate experiment to be done.

I asked him if anyone really worked on this and he answered: "a few, not many, but you could do worse with your life than get interested in this; I think that if you are interested at all you should go and see Josh Lederberg." I did go to see the great Lederberg, who was a legend in his time. And this was the next time I met a visionary, because he was more romantic about it than Horowitz. He said there was a mission coming up which is a very important mission. . . . we are talking about sending a spacecraft to Mars with a parachute to land on its surface.

After I got home, three or four weeks passed before my colleagues said, "You're driving us crazy, you've talked about nothing but this mission to Mars and the kind of experiments. Why don't you go and do something about it instead of talking about it?" A few months later I accepted a job at JPL.

I doubt that as a young person I would have accepted a 1975 landing back in 1961. But I was constantly drawn in. At that time we were talking about a 1967 mission, then a 1969 mission, and bit by bit the time passed. The high point was when Voyager was conceived. It brought many more biologists into the program. But Voyager was wrong; it may have been attractive, but it was not responsive to the needs. Viking was, in a sense, a phoenix that rose from the ashes of Voyager.

I believe that Viking's main goal is biology. It isn't the only goal, but it is probably one of the only times in our lifetime that we will get a mission of this complexity to Mars.

It suffers because we are not sure yet which experiments are going to work. I would be supremely happy if one of them worked . . . gave the first clue that says yes or no. It is very hard to disprove that there is life on Mars, but you can prove it. You cannot disprove it until you have turned over every rock and tried every site. We tend to think of Mars as being a place, but . . . it is a world. There are many, many places there.

The People and the Plan

Chapter **4**

Expedition to the Red Planet

As early as 1925, Walter Hohmann had worked out the general route that Viking would take on its journey to Mars. To carry the maximum payload of scientific instruments, spacecraft flying to Mars have to follow a path that takes them approximately halfway around the sun in a way that grazes the orbits of earth and Mars. This is close to a "Hohmann ellipse," a type of trajectory between planetary orbits that was described by Hohmann in *Die Erreichbarkeit der Himmelskorper* (The attainability of the heavenly bodies). Hohmann discussed the landing of a spacecraft on Mars after a journey of 265 days, and he calculated all the changes in velocity needed to escape from earth, travel out from earth's orbit to Mars' orbit, and match speed with and land on Mars. He also calculated the size of rocket vehicle needed for such a journey, based on expected exhaust velocities for an advanced rocket engine.

Viking was planned as such a mission; two identical unmanned spacecraft were scheduled for launching about 10 days apart during August 1975. Each spacecraft, consisting of an orbiter and a lander, was to be lifted from earth by a Titan/Centaur booster and sent onto a trajectory that would carry it along an elliptical orbit to the orbit of Mars. It would rendezvous with Mars and be placed in orbit around it. After several weeks of orbiting, during which preselected landing

sites were inspected and a suitable site chosen, the lander spacecraft would separate from the orbiter and go down to the Martian surface.

The Titan/Centaur booster consisted of a two-stage, liquid-propellant rocket, developed from the Titan ICBM, plus two large solid-propellant rockets (figure 4.1). The whole was topped by a high-performance Centaur upper stage. At lift-off the solid-propellant rockets would provide the enormous thrust needed to accelerate the combination of liquid-propellant stages and spacecraft. When the solid rockets had burned all their propellants, they would be jettisoned, and the first liquid stage of Titan would fire, followed by its second liquid stage. When all the Titan propellants had been consumed, the spent shell of engines and propellant tanks would separate from the Centaur uppermost stage, which then would thrust itself and the spacecraft into orbit around the earth.

After coasting for 30 minutes in that orbit, the Centaur would reach a position from which it could enter the path to Mars. The Centaur engines would restart and burn the remainder of the propellants, breaking orbit and accelerating the spacecraft to the right speed for its journey outward from the sun. The spacecraft (figure 4.2) then was to separate from the

4.1 The Titan/Centaur booster used to launch the Viking spacecraft to Mars, is here shown in cross section. The two-stage, liquid-propellant vehicle is flanked by two solid-propellant modules and topped by the high-performance Centaur upper stage. The Viking spacecraft fits snugly into the nose fairing at the top of the big rocket.

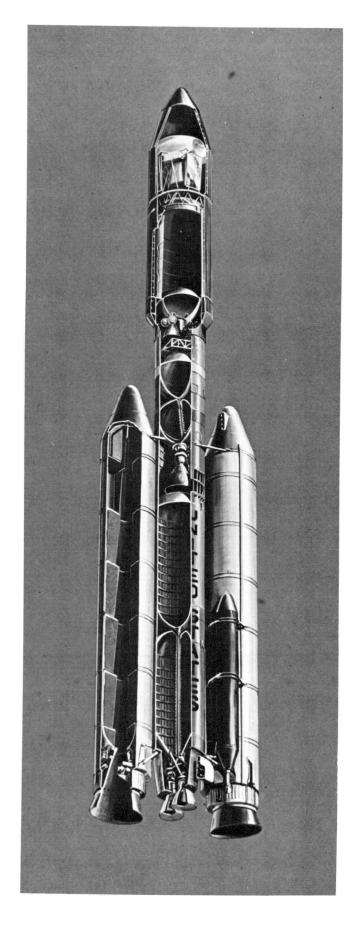

Centaur, whose spent shell would later be deflected by thrusters so that it could not crash onto Mars.

Viking was expected to make several flight corrections during its journey, based on navigation information acquired from earth-based tracking stations of NASA's Deep Space Network. The network to support the mission consisted of three stations, located in California, Australia, and Spain. Each station had a 210-foot (64-m) diameter antenna and an 85-foot (26-m) antenna.

In addition to tracking the precise path of the spacecraft, this system would process three kinds of data: engineering telemetry to tell engineers here on earth how the various pieces of equipment within all four spacecraft were functioning, measurements from the science instruments, and commands to the spacecraft to start or change operations at the spacecraft, such as switching equipment or scientific experiments on and off and directing the cameras how and when to take pictures. The center for communicating with the spacecraft was to be located at the Jet Propulsion Laboratory.

Communications with Viking would take longer and longer as the spacecraft traveled farther from the earth. When it reached Mars, a one-way message would take 20 minutes to flash across the void at 186,000 miles a second. This would mean that a round-trip minimum of 40 minutes must pass before a command from earth could be received by the spacecraft and its response could get back to the earth. For this reason, automation was essential. Operations that could not be interrupted, such as the soft landing on Mars, had to be performed automatically by an on-board preprogrammed computer.

Power for the spacecraft's electrical system would come from solar panels that opened on command when the Viking was placed on its interplanetary

Expedition to the Red Planet

4.2 The top picture shows how the lander, enclosed in its bulbous bioshield, was attached to the orbiter spacecraft. The bottom picture shows the uncovered lander in a simulated Martian landscape.

path. The power output from the solar cells was to be supplemented by batteries to meet peak loads. Most of the time these batteries would not be needed to supply power and would be kept fully charged by the solar panels. Small attitude-control jets were mounted at the tips of the solar panels to keep the spacecraft stabilized and correctly oriented in space.

The oribiter also would supply power to the lander spacecraft which it carried, so that the lander could be checked periodically during the long voyage to Mars. The lander itself would carry radioisotope thermoelectric generators (RTG's for short) that convert heat from a nuclear source into electricity to supply electrical power when the lander was on the surface of Mars. It could not use solar cells because they might be damaged or made useless by the Martian dust storms. It, too, had batteries to meet any power demands above the capacity provided on a continuous basis by the RTG's.

As the spacecraft neared Mars, it would be maneuvered into the proper orientation to enter an elliptical orbit around the planet. A rocket engine carried by the orbiter provided sufficient thrust to place the combined orbiter and lander into orbit. This orbit was matched to the rotation period of Mars so that the spacecraft would circle the planet once during each Martian day, or "sol," with the lowest part of the orbit being over the chosen landing site.

The orbiter's first task was to survey and certify a landing site. Then the lander would be made ready for separation and landing. It would be sent a series of commands to store in its computer. These commands would operate all the systems of the lander during its descent to Mars, and would include a program to operate for 90 days on the surface and conduct its experiments. This program made it possible for the spacecraft to act independently of earth and to complete its nominal mission of 90 days even if it could not receive any more commands.

The lander was contained in a protective capsule called an aeroshell, to protect it from the intense heat generated when the spacecraft plunged at high speed into the Martian atmosphere. During descent and landing the lander would maintain radio communication with the orbiter, which would act as a relay station between the lander and the earth.

Atmospheric drag on the aeroshell would slow the lander until, at about 21,000 feet (6,400 m) above the surface of Mars, it was moving slowly enough to use a parachute. Then the aeroshell would be jettisoned. The parachute would continue to slow the descending spacecraft until it was about 1 mile (1.6 km) above the surface. Then three rocket engines on the lander would ignite, the parachute would be jettisoned, and within 13 minutes the lander would touch down on the surface of Mars.

That was the plan, but much development work had to be completed before it could be realized—development of the two spacecraft, of the scientific instruments they were to carry, and of a system of mission operations to control the mission remotely from earth.

Viking was in the tradition of earlier expeditions of exploration on earth. But earth was explored by people physically traveling, and it took many months, sometimes years, before the exploration could be shared with the general public. By contrast, Viking's interplanetary expedition took place with remotely controlled machines, and through radio communications the public shared in the discoveries as rapidly as they were made.

To find out about the design and development of the landers, I talked with John Goodlette, the chief engineer for the Viking program at Martin Marietta. Having participated in planning for a Voyager spacecraft through company-funded studies, he became involved in Viking when the Voyager program was canceled by Congress in 1967. Two approaches were in vogue for landing on Mars, explained Mr. Goodlette. One was to make a direct flight entry of the kind used by the Russians, in which the lander separated from the orbiter before the latter went into orbit. But the Russians found that this was not a good procedure since it permitted no flexibility in the mission. Their spacecraft arrived at Mars during a major dust storm and had no choice but to plunge into the storm. They did not survive. Goodlette and his group preferred the alternative approach of carrying the lander into orbit and conducting some orbital reconnaissance before despatching the lander to the surface. NASA also favored this approach, and it was chosen for Viking.

Several technical aspects of spacecraft design needed improvement before the Mars mission could take place, said Goodlette. The most important was a way to sterilize the lander and yet preserve high reliability of its components. The question of sterilization went back to the early days of the space program. At that time several biologists and other scientists, including Dr. Joshua Lederberg, had successfully pushed for international agreements to sterilize all spacecraft that were intended to land on planets. This sterilization program attempted to ensure that if space missions did ultimately find life on another planet there could be no doubt that it was local and not a stowaway from earth. The orbiters did not require sterilization because they would not land on Mars, but the landers had to be sterilized.

Heat sterilization had been selected for Voyager, and it was continued for Viking. In this method the complete spacecraft is heated to 235 degrees Fahrenheit (112°C), a temperature at which microbes are killed, and is kept there for almost 2 days. To prevent contamination after sterilization, the spacecraft is sealed in a container. It is analogous to preserving fruit by heating and sealing it in a bottle. Unfortunately, electronic and other components of the spacecraft, such as computers, are not normally designed for steriliza-

tion temperatures. There were also other important aspects of using this kind of sterilization before launch. What would happen if a fault developed between sterilization and launch? How could people work on the spacecraft without having to resterilize it?

After years of study and testing, Martin Marietta selected the component parts needed to make the spacecraft suitable for heat sterilization. The parts chosen were not only able to survive the heating but also retained high reliability for the long space voyage. The spacecraft was built in a scrupulously clean environment. All parts were cleaned before assembly to keep microbial contamination as low as possible. Before it was sterilized, the lander was sealed inside a lightweight bioshield made of woven fiberglass on an aluminum structure. It was effectively inside a preservation bottle. There it remained throughout the countdown and the launch, until it was above earth's atmosphere and had separated from the launch vehicle on its way to Mars.

Another task the engineers faced, said Mr. Goodlette, was to design the lander so that it would consume minimal amounts of electrical energy because the two generators it carried could provide a total of only 75 watts of power continuously. Viking became a model of energy conservation. All the activities of the complicated machine on the surface of Mars had to take place with the amount of energy consumed by a typical desk lamp. Anything not in use was automatically switched off by the lander's computer. But this meant that the computer also had to turn everything on again when needed, which called for added capability. In fact, the design of the very advanced computer needed to operate the lander for 90 days, independent of control from earth, was a major achievement in itself.

The lander had to be tough. It had to withstand the launch and then the long journey to Mars lying quiescent beneath the orbiter. It had to survive the enormous temperature and deceleration forces of entry into the Martian atmosphere, and also had to operate on the surface of Mars for at least 90 days in temperatures that might vary by as much as 160 degrees Fahrenheit (90°C) between day and night. Moreover, the lander had to be capable of withstanding dust storms that were expected to sweep across the surface of Mars during the mission.

Special thermal coatings were developed to protect the lander. Individual components were arranged within the spacecraft so that less vital components protected vital ones. The most sensitive pieces of equipment had thermal heaters to warm them should temperatures drop too low during the frigid Martian nights. Also, some of the waste heat from the RTG's (mounted outside the main body of the lander) could be directed into it when required.

A vital part of the mission was collection of samples from the surface of Mars and placement of these samples into hoppers for several of the experiments. The biology experiments and the analysis of surface

material could not take place without these samples. A powered collector scoop was designed that had a small head like a shovel but consumed only 15 watts of electrical power. A flexible boom pushed the scoop out 10 feet (3 m) from the spacecraft and then brought it back to dump the soil into hoppers that projected on top of the spacecraft's body.

Another task was to design rocket engines for the lander that would not disturb the soil at the landing site too much. Each lander needed three rocket engines to stabilize its descent, like legs of a tripod, and get it safely down on Mars. They were tested in simulated landings, but their exhaust jets dug holes in the soil surface and sprayed soil in every direction. This would have been confusing to the soil sampling experiments, which ideally needed an undisturbed surface. The solution was to redesign each rocket with an array of 18 small nozzles instead of a single large one (figure 4.3). This effectively spread the exhaust and considerably reduced the disturbance to the soil, yet it did not affect the thrust of each rocket too much.

Another major problem faced the Martian engineers: delivering the landers in time for the launch window while also thoroughly testing each spacecraft at every stage of its progress, from construction to the launch. A completely automatic, computer-controlled checkout system was developed for Viking. This test program was specially designed to show trends and to issue warnings if any part of the spacecraft did not perform as expected. It proved to be a major confidence builder as launch day approached and fewer and fewer warnings were given.

4.3 Dr. James Fletcher (*at right*), Administrator of NASA, inspects the multiple-nozzle rocket engines used on the Viking landers to prevent disturbance to the landing site as the spacecraft touched down on Mars.

The orbiter spacecraft also presented several technical challenges. At the Jet Propulsion Laboratory, design and fabrication of the orbiter came under the management of Henry Norris, who had been involved in the development of several Mariner spacecraft. The Laboratory had, indeed, proposed a 1971 Mariner mission to send a probe into the Martian atmosphere using an aeroshell and a parachute, and it had also been involved in studies of Voyager-type spacecraft. Viking, however, had to be bigger than any Mariner (although smaller than a Voyager spacecraft) in order to carry the lander to Mars.

The biggest change in early plans for Viking came in January 1970, when the launch date was moved from 1973 to 1975 and development funding was limited for two whole years. The number of people working on the project had to be cut back. One hundred and fifty people had to find jobs elsewhere in a short period during December 1969 and January 1970. Fortunately most of these people were able to transfer to the Mariner Venus/Mercury program, which meant that they were available for transfer back to Viking when it was restored to activity two years later. Otherwise they probably would have been lost to the Viking program, and the mission would have been hard put to make the scheduled launch date.

Although these cutbacks were traumatic experiences to the individual scientists and engineers, the delay to 1975 permitted a useful reappraisal of the reliability of the mission in terms of redundancy of critical components of the spacecraft. Redundancy is an aerospace term for carrying spare parts in the spacecraft that can be brought into operation during a mission, either automatically or by command, to replace an ailing or a failed component. The need for such spares was greater for a 1975 mission than for one in 1973. A 1973 mission could use a path to Mars that took only 3 or 4 months; by contrast, a launch in 1975, at a different configuration of the planets, required 10 months of travel to reach Mars. The 1975 mission thus placed a premium on reliability for success of the mission. A major decision to carry more hardware on the spacecraft to achieve redundancy of vital components was made in October 1971.

The rocket engine for the orbiter, although basically the same as that used in Mariner 9, needed higher efficiency to develop more thrust. Also it had to fire for a longer time, which meant that the combustion chamber and nozzle had to be cooled. These improvements would allow the rocket engine to place a heavier spacecraft and the lander into orbit about Mars. For the longer period of thrust, additional propellants had to be carried, and therefore a bigger tank than Mariner had had. Because the engine had to be fired several times, it needed a more complicated valve control system that used both pyrotechnic valves (one-shot explosive devices) and electrically operated valves.

The Viking orbiter consumed more electrical energy than did the earlier Mariners. This energy was provided by a greater area of solar panels, to provide 620 watts continuously. The panels were so big that they had to be folded for the launch and extended after the spacecraft separated from the Centaur. When this amount of power was insufficient for peak loads, it was supplemented by two 30-ampere-hour nickel-cadmium storage batteries.

Both the orbiters and the landers carried scientific experiments. Each orbiter's experiments were mounted on a special scan platform that could be controlled separately from the spacecraft (figures 4.4 and 4.5). In this way the spacecraft, orbiting Mars, could be maintained in a constant orientation to the sun to obtain power through the solar panels, and to the earth to return data, while at the same time the scientific instruments could be directed to specific targets on Mars.

The major experiment carried by the orbiter was an imaging system to photograph the surface of the planet in greater detail than had been possible with Mariner 9 when it orbited Mars in 1971. This imaging system consisted of two identical cameras, each made up of a high-powered telescope, filters, a TV camera tube, and associated electronics. From an altitude of 930 miles (1,500 km), these cameras were able to reveal anything on the Martian surface that was about the size of a football stadium or larger.

On the same scan platform as the cameras was an infrared spectrometer to detect and map the extent of water vapor in the Martian atmosphere. Nearby, the infrared radiometer was to measure the intensity of radiation from Mars and thus determine the temperature of the planet's surface.

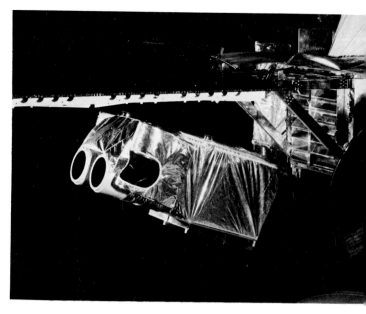

4.5 The cameras and infrared instruments carried by the orbiters were mounted on a movable scan platform so they could be directed toward targets on Mars without changing the orientation of the spacecraft itself.

For each scientific experiment there was a team of scientists with a team leader. Michael Carr of the U.S. Geological Survey in Menlo Park, California, was leader of the imaging team for the orbiters. The first task of Dr. Carr's team when Viking 1 arrived in orbit around Mars was to obtain the best possible images to approve, or certify, the landing site for the first lander. Actually the orbiter would also photograph Mars while approaching the planet, and would obtain colored pictures surpassing anything available from earth. Since these approach pictures would be the first to show the whole globe of Mars in detail since 1969 (the approach of Mariner 9 had not produced useful pictures because of the dust storm), the sequence was of great interest to planetologists.

After the landing site was certified and the first lander was safely down on the planet, the orbiter would then concentrate on investigating the geology of many interesting areas of Mars, gradually extending detailed coverage over most of the planet's surface.

Of major importance, said Dr. Carr, would be an attempt to find out how the eroded channels, first discovered in the Mariner 9 photographs, were formed. Other sequences of pictures from orbit would be

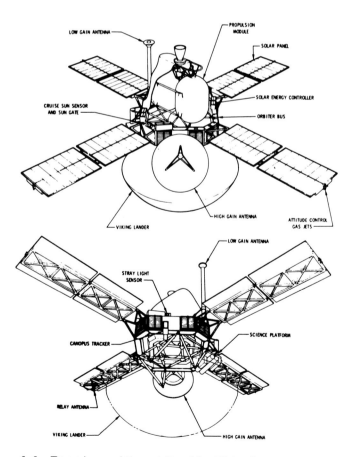

4.4 Two views of the orbiter, identifying its main components.

aimed at a detailed survey of the great canyons of Mars, the chaotic terrain, and the calderas and slopes of the huge Martian volcanoes. As the mission progressed, more pictures would be obtained of the polar regions which, as Mariner 9 had revealed, contained some intriguing, unexplained layering features. The orbiter would also look at changes in surface features during the Viking mission, and at the beginning and development of dust storms.

Hugh Kieffer, of the University of California at Los Angeles, was team leader for the thermal mapping experiments using an infrared radiometer carried by each orbiter. Dr. Kieffer was also involved with site certification, because infrared observations from orbit could provide information about the nature of the surface. Large boulders, for example, have a much different diurnal (daily) temperature variation than does sand. After the landing site was certified, the instrument would be used to determine the temperature of the Martian atmosphere as well as the surface.

One aspect of the experiment, said Dr. Kieffer, would be to find out whether Mars had cellular convection cells of solar-heated, rising columns of air as does the earth. Stability of the atmosphere could be determined from orbit, too. This was important to know before the lander descended; a major instability in the atmosphere over the landing site could lead to delay in the landing because it could upset the stability of the lander, as turbulence affects an aircraft.

An exciting potential of the radiometer was determining the temperature of the polar caps to resolve the question of whether the residual caps were water ice.

The team leader for the third orbital experiment was Barney Farmer of the Jet Propulsion Laboratory. Dr. Farmer had developed a model for the behavior of water vapor on the surface of Mars and expected the infrared water-vapor detector to confirm the model. All evidence pointed to water vapor being restricted to a layer close to the Martian surface, below 3,000 feet (1,000 m). The big question was, Where did this water go if the amount in the atmosphere varied each day as expected? Did it fall from the atmosphere as ice crystals? Or did it become absorbed in the soil? Dr. Farmer hoped to resolve such questions with data from the atmospheric water detector.

There are at least two types of water clouds on Mars, he pointed out. Morning and evening clouds are probably surface hazes caused by changes in temperature between day and night. They are similar to the morning mists often seen in low-lying fields on earth. The other clouds are caused by upward motion of a mass of air to elevations at which condensation occurs because of colder temperatures. These clouds are seen each day on the flank of the Martian volcanoes.

Dr. Farmer suggested that ice crystals might form on dust grains in the air late each day and fall to the surface of Mars, remaining as ice overnight. When the temperature rose each day after sunrise, such ice and dust could go through a liquid phase before the water evaporated into vapor. In some latitudes of Mars there might be a daily "wet" phase on the surface which lasted for a few hours, and this would be important to any Martian life forms. An expected result of the Viking mission, said Dr. Farmer, would be to establish how the amount of water varied on a daily, annual, and perhaps epochal basis.

While the orbiter's scientific experiments extended the observations made by Mariner 9, the lander's experiments would explore entirely new fields, at least insofar as the American space effort was concerned. The lander's scientific experiments comprised two major groups that were quite separate even though

Expedition to the Red Planet

many were interrelated. The first group of experiments—entry science—would occupy the period of passage through the atmosphere of Mars down to the surface. The second group—landed science—would apply when the lander was safely on the surface.

The team leader for entry science was Alfred Nier of the University of Minnesota. The experiments used a variety of instruments carried on the lander and its protective aeroshell (figure 4.6).

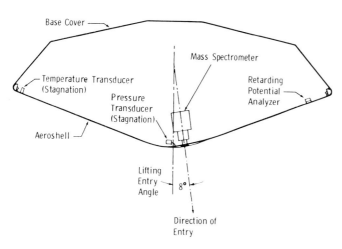

4.6 The aeroshell that protected the Viking lander during its high-speed entry into the Martian atmosphere carried several experiments to measure the characteristics of the atmosphere.

In the highest regions of the Martian atmosphere, explained Dr. Nier, the entry experiments would look at the properties of the ionosphere, where atoms and molecules of atmospheric gases become electrically charged by incoming solar radiation. Earlier the Mariners had passed radio waves through the upper atmosphere of Mars and found that there existed a Martian ionosphere similar to earth's D and E ionized

layers at 50 and 60 miles (80 and 100 km) respectively, but at an altitude of about 80 miles (130 km) above Mars' surface. Mounted in the protective aeroshell of each lander would be a retarding potential analyzer. It would apply different voltages to measure the concentration and charge of ions and the concentration of electrons in the rarefied upper atmosphere. The experiment was also expected to reveal details of how the solar wind (a stream of electrons and protons from the sun) interacts with the ionosphere of Mars. Mars is, of course, not protected from the solar wind, as earth is, because Mars does not have a strong magnetic field like the earth's. Earth's field deflects the solar wind around us.

As the spacecraft plunged deeper into the Martian atmosphere and pressure increased, the retarding potential analyzer could no longer operate. Then the atmosphere would be investigated by an upper-atmosphere mass spectrometer, also carried in the aeroshell. This instrument would determine the constituents of the uncharged (electrically neutral) atmosphere. It could detect not only gases such as argon, nitrogen, oxygen, and the expected carbon dioxide and carbon monoxide, but also the various isotopes of these gases. This was very important for finding out how gases were released from the interior of the planet to give rise to its first atmosphere, and how that early primitive atmosphere changed to the Martian atmosphere of today. Such information is needed to understand how other planets, including earth, developed atmospheres, and how they changed with time as the planets evolved in their separate ways.

The upper-atmosphere mass spectrometer could easily detect several percent of nitrogen if it was present in the Martian atmosphere, and also any oxygen there. Perhaps the most important measurement would be of argon. A search for this gas was to be given top priority in the processing of data from the entry experiments. The urgency stemmed from some results derived from the Soviet attempts to land on Mars at the 1973 opportunity.

At the 1973 launch window the Russians sent four spacecraft to Mars. Mars 4 and 5 were designed to orbit the planet and were launched July 21 and July 25 respectively. In February 1974 the first of these spacecraft reached Mars. Although it returned some pictures, it failed to enter into orbit around Mars and flew by the planet. Mars 5 also arrived at Mars in February and successfully entered orbit. It returned pictures for nearly 2 weeks before running out of attitude-control gas.

Mars 6 and 7 were designed to fly by Mars and release capsules to land on the planet's surface. Mars 6 was launched on August 5 and Mars 7 on August 9, 1973. Both reached Mars successfully, made a successful flyby in March 1974, and released their capsules. The capsule from Mars 6 entered the atmosphere and landed at about 24 degrees south and 24 degrees west, on the southern boundary of a dark area called Margaritifer Sinus. Some information was sent back regarding the atmosphere during entry but, as expected, the spacecraft stopped transmitting data just before touchdown. When the data flow should have started up again after the landing there was only silence. The capsule from Mars 7 did not separate correctly and missed the planet entirely.

An important indirect scientific result was obtained from the Mars 6 capsule that landed. It carried a mass spectrometer to identify atmospheric gases. This instrument was pumped to a low internal pressure during the descent in preparation for its operation on the surface after landing. However, the pressure inside the instrument did not go down as expected. Consequently the Russian scientists concluded that there must be at least 20 percent argon in the Martian atmosphere to account for the anomalous behavior of the mass spectrometer.

This announcement was important to Viking because one of the major experiments of the lander was a search for organic molecules on the Martian surface. Such molecules could be the dead "bodies" of microorganisms, or even prebiological building blocks for a biota that had not yet evolved on Mars. The instrument used a mass spectrometer to analyze gases released from soil samples by heating. This mass spectrometer also was to be used in a separate series of experiments to sniff the atmosphere at the surface and find its composition, thereby supplementing the entry experiment by providing several analyses of the atmosphere during the period when the lander was active on the surface.

If it proved true that there was 20 percent or more argon in the atmosphere, the lander experiment could not take place in the sequence originally intended. Atmospheric analysis would have to wait until after the soil analyses had been successfully completed. Analysis of the soil was more important because a crucial part of the search for life on Mars was to try to find organic molecules in the soil. Argon in the atmosphere, if taken into the instrument, could ruin it and prevent subsequent analyses in search of organic particles in the soil. However, if the soil analyses had to be made first, later atmospheric studies might be contaminated by the residues from the soil analyses.

The entry experiments included sensors in the aeroshell and the lander to measure temperature and pressure from the top of the Martian atmosphere down to the planet's surface. Also, the lander's accelerometers, used to supply information to its computers to aid a safe landing, were available to help determine the density of the Martian atmosphere and how it varied with height above the surface.

For the first time, a complete profile of the atmosphere of Mars could be obtained by measurement

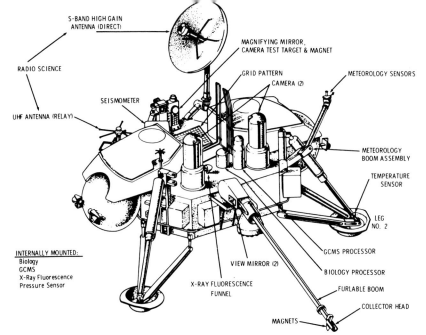

S-BAND HIGH GAIN
ANTENNA (DIRECT)

MAGNIFYING MIRROR,
CAMERA TEST TARGET & MAGNET

RADIO SCIENCE

GRID PATTERN
CAMERA (2)

METEOROLOGY SENSORS

SEISMOMETER

UHF ANTENNA (RELAY)

METEOROLOGY
BOOM ASSEMBLY

TEMPERATURE
SENSOR

LEG
NO. 2

INTERNALLY MOUNTED:
Biology
GCMS
X-Ray Fluorescence
Pressure Sensor

GCMS PROCESSOR

VIEW MIRROR (2)

BIOLOGY PROCESSOR

X-RAY FLUORESCENCE
FUNNEL

FURLABLE BOOM

COLLECTOR HEAD

MAGNETS

within the atmosphere. If both spacecraft successfully landed on Mars, these atmospheric profiles would be made at two different latitudes, one at a low and the other at a high northern latitude. The two landers also carried radar altimeters to aid their descent, data from which could be used to help establish atmospheric conditions and the profile of the Martian surface below the path of the lander.

On the surface, experiments would include two types—those that depended on a sample of Martian soil, and those that were independent of soil sampling (figure 4.7). One of the experiments that did not need a soil sample was meteorology. The leader of the science team for this experiment was Seymour Hess of Florida State University in Tallahassee. Dr. Hess described the meteorology experiment as "a net stretched in time rather than space" to collect information about weather conditions on the surface of Mars. On earth, meteorologists have the advantage of spreading their observing stations widely over the planet, but successful landings of two Vikings on Mars would provide only two weather stations.

If both landers survived through a Martian year, however, they could provide information on daily and seasonal variations of the weather there. They could monitor such things as the passage of cold fronts and the onset and progress of dust storms.

The two landing sites provisionally chosen before the mission were almost ideal from the standpoint of meteorology, said Dr. Hess. Neither was on the equator,

4.7 This drawing shows the location of the various experiments on the Viking lander in its landed-science configuration.

both were in the same hemisphere and reasonably close to each other, and both were on reasonably flat surfaces away from mountains.

Dr. Hess pointed out, too, that Mars is somewhat similar to the earth in terms of atmospheric dynamics, but it is simpler because there is so much less water on Mars. The evaporation and precipitation of enormous quantities of water on earth considerably complicate terrestrial meteorology. He expected to be able to predict weather conditions on Mars more easily than we can on earth. He also expected to be able to detect a solar tide in the Martian atmosphere, owing to solar heating that would cause the air to rise. This should appear as a pressure change each day at the landing sites. Winds at the landing sites should also vary in response to this solar tide as part of the global oscillations of wind patterns.

The meteorology experiment used a group of sensors mounted on a boom that was to be extended from the lander after touchdown. Over each day and night these sensors would telemeter to earth the temperature and pressure of the atmosphere at the landing sites and the direction and velocity of the Martian winds.

Another ground experiment that did not rely on obtaining soil samples was the seismology experiment, for which the team leader was Don Anderson of the

Expedition to the Red Planet

California Institute of Technology. Dr. Anderson described the experiment as essentially trying to find out if Mars is still active today. Major faults that are one evidence of tectonics, or crustal movement, appear nearly everywhere on the surface, and the big volcanoes look quite young. Major questions facing geologists are, How young? and even more important, Are these volcanoes still active? The Tharsis area on which the big volcanoes are located is uplifted high above the mean level of the planet's surface. This must give rise to stresses in the crust of Mars. Scientists were interested in finding out how these stresses are relieved.

The heart of the seismometer package was a three-axis seismometer to measure vibrations in the surface along three directions perpendicular to one another. Such vibrations would cause a spring-supported pendulum to move and generate in a coil small electrical currents that were proportional to the amount of movement. Unfortunately the only place where this instrument could be mounted was on top of the Viking lander. As a result the seismometer would be affected by vibrations resulting from use of the other equipment and from the lander's being shaken by Martian winds.

His team's initial task on the Martian surface, said Dr. Anderson, would be to establish the characteristics of the unwanted vibrations. After the background was known, the seismometer could then differentiate between these vibrations and seismic activity in the crust of Mars.

There was no way of predicting in advance what the seismic activity of Mars might be. A similar piece of equipment on earth, say in California, might have to wait for a month before it registered a minor earthquake. So this was a long-term experiment. It was the only experiment that depended for its full success on there being two landers safely down on Mars. To

ascertain direction, distance, and depth of marsquakes, the experimenters needed two seismometers oprating at different points of the surface of the planet. The seismology experiment was also the only Viking experiment that was directly concerned with the interior of the planet, although the radio science experiment used tracking data to infer what the interior was like by analyzing the planet's effects on the path of the orbiting spacecraft.

Another important surface experiment that was independent of soil sampling was lander imaging, or photography. This made use of facsimile cameras that produce a picture as a series of lines side by side (figure 4.8). The cameras were mounted on top of the lander. Developed specially for Viking, they required several engineering breakthroughs. A nodding mirror sequentially scanned the scene in front of a camera as a series of vertical strips arranged side by side until a complete panorama was covered.

Each camera took 20 minutes to scan a full scene. The camera did not require a lens that might be scoured by Martian dust storms, but showed objects in sharp focus from close to the lander to the far horizon. The viewing window was protected by a cover when not in use. The cameras could take photographs in black and white, full color, and infrared. Using the two cameras of a spacecraft together produced stereo pairs so that the view in front of the lander could be seen three-dimensionally.

Dr. Thomas Mutch, of Brown University at Providence, Rhode Island, led the lander imaging team. He described the instrument as a 'jack-of-all-trades." It would be used to certify the locations from which samples of Martian soil would be scooped up; it would show the progress of soil sampling; and it would check the interactions of the lander's footpads with the Martian soil and how this soil was disturbed by the lander. It would survey the whole area around

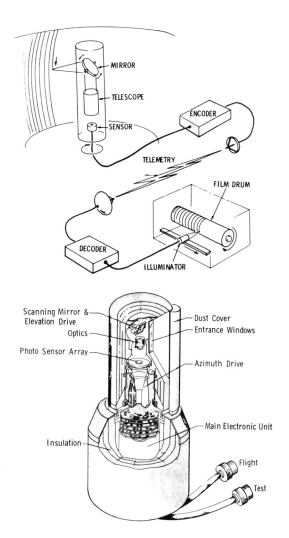

4.8 One of the most important experiments on the lander was its imaging system, which would provide the view of the surface at each landing site. Each lander carried two cameras. The system built up an image as a series of vertical strips, as shown in the top diagram. The camera itself is shown partly cut away in the bottom diagram.

Experiments to determine the physical and magnetic properties of the Martian surface did not have specific experimental packages. Instead they were planned to use data returned by the other experiments, and particularly by the cameras, which would show how the surface was affected by the landing and by subsequent activities of the lander, including use of the soil sampling arm.

Apart from photographs of the surface of Mars, the Viking activities that stimulated the most public interest were undoubtedly those connected with sampling and analysis of the surface materials, particularly whether or not the soil of Mars contained living things. This group of experiments relied upon Martian soil being collected and deposited into the hoppers of several scientific experiments.

the lander in great detail and in color at different times of day and throughout the mission, looking for evidence of living things, for changes, and for information about the nature of the Martian surface.

Through the eyes of these cameras, people on earth would see Mars as though they were actually on the surface of the planet, sitting on the lander and looking around them.

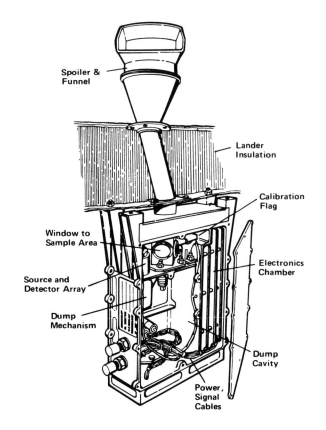

4.9 An X-ray fluorescence spectrometer, shown in cutaway in this drawing, analyzed the elemental composition of the Martian surface.

The inorganic analysis experiment was designed to find out which elements are present in the Martian surface material (figure 4.9). Soil samples would be irradiated with X-rays generated by a source within the spacecraft. The soil particles would then produce secondary X-rays characteristic of elements in the soil sample. The team leader was Priestley Toulmin III, of the U.S. Geological Survey in Reston, Virginia. Dr. Toulmin explained that the experiment had several objectives. Its data would help us understand the kinds of primary rock materials on the surface of Mars and the secondary "weathering" processes that redistributed these materials. In addition, the experiment was expected to provide information about minerals.

An extremely important experiment from the stand point of biology, though not classed as a biology experiment, was developed by a team led by Klaus Biemann of the Massachusetts Institute of Technology. This experiment used a gas chromatograph and mass spectrometer, which would separate different compounds carried by a flow of gas and then analyze these compounds from the Martian soil for traces of organic molecules. As described earlier, the mass spectrometer was also to be used to find the composition of the atmosphere. The instrument became widely known by the initials GCMS.

The original concept for an organic analysis experiment was developed at the Jet Propulsion Laboratory for the Voyager program. It was an ambitious project to shrink a whole laboratory worth of equipment into a volume of about a cubic foot. The task of doing this for Viking fell to Litton Industries, assisted by Perkin Elmer, which made the mass spectrometer, and Beckman Instruments, which made the gas chromatograph (figure 4.10). When the whole package was almost ready, troubles developed in the soil processor that ground soil particles to a fineness needed for the experiment. The soil particles acted as abrasives and

(PHOTO, LITTON GUIDANCE & CONTROL SYSTEMS)

4.10 The gas chromatograph–mass spectrometer, shown here, proved to be one of the most critical instruments in the search for life on Mars. Its purpose was to seek organic molecules in the Martian soil.

jammed the transporter that moved soil from the grinder to the test chamber. Another problem was that metal parts from the grinder became mixed with the soil and produced erroneous readings. Dr. Biemann made the eleventh-hour decision to fly one unit as it was, while the unit for the other Viking lander was modified to use hardened material for the soil grinder. The schedule was met and the two units were fitted into the landers in time for the launches.

As mentioned earlier, a serious operational problem faced the experimenters. The original sequence was to allow the mass spectrometer of the first lander to be used immediately after landing to sample the atmosphere at the landing site. The operational plan

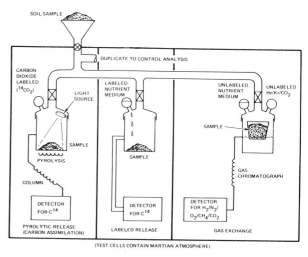

(PHOTO, TRW SYSTEMS)

4.11 The biology package of the Viking landers consisted of three separate experiments in one unit. These experiments are shown in simplified form in this drawing.

did not anticipate having soil samples before the eighth day on Mars. Analysis of the soil could contaminate the apparatus and affect the analysis of the atmosphere, so ideally the atmospheric analysis should be carried out first. Then the gas chromatograph would receive a soil sample and pass the volatile components of the sample to the mass spectrometer to identify the organic compounds.

The large amounts of argon indicated by the Soviet results would damage the mass spectrometer. A decision was made to delay the atmospheric analysis until after the soil analysis if argon was confirmed in the entry experiments. This was the instruction loaded into the computer memory of the lander of Viking 1 before the landing. However, if the entry science experiments showed only a small percentage of argon, commands would be sent to the lander after the landing to change the sequence.

In that case, explained Dr. Biemann, the first operations on the surface would be to confirm that the complex instrument had survived the landing. Second, if there was a lot of organic material on Mars (which he regarded as unlikely), a strong positive signal would be expected. If not, much analysis would be required. Then he expected to be able to identify the major components of the organic molecules. But he emphasized that the experiment was not one of life detection, even though it had important implications about life on Mars.

Three biological experiments were packed into each Viking lander (figure 4.11), as devised by Oyama, Horowitz, and Levin. Each was an independent experiment, making a different assumption about what Martian life forms might be like. All three searched, however, for microbial life. These three experiments are described in more detail in chapter 7.

Dr. Klein, who had responsibility for all the biology experiments on Viking, described the philosophy be-

hind the experiments and how they had evolved. When the biology team was first organized and started to think about looking for life on Mars, the members quickly concluded that any single experiment that could be devised would be an extremely long shot. No one knew whether or not there was life on Mars, nor did anyone know the principles of life well enough to say on what life would be based. But they all agreed that a good case could be made, based on cosmochemistry and chemical evolution, that if there were life on Mars it would be based on carbon as the major backbone, and on hydrogen, oxygen, and nitrogen as being the most prevalent elements in stars and in the solar system.

Experiments in laboratories had already demonstrated that very simple compounds, such as methane (CH_4) and ammonia (NH_3), could readily be synthesized into precursors of life. It was easy to conclude that whatever was done to seek life on Mars, the smart thing the first time around would not be to talk about "silicon bugs" or "titanium bugs," for which there was no philosophical or scientific background, but to assume that carbon life is normal in the solar system. Carbon compounds had been found in meteorites, and it seemed reasonable that they would also exist on Mars.

Expedition to the Red Planet

In May 1970, TRW Systems was awarded a subcontract by Martin Marietta to build a set of laboratory models of four biology experiments—those of Horowitz, Levin, Oyama, and Vishniac. During the following year the development, design, and testing of the units that would be flown to Mars started. Budgetary constraints on Viking's hardware program made it necessary to drop the experiment by Vishniac.

The delivery of soil to the individual experiments caused design problems. The soil had to be divided into small samples and distributed. But the particles jammed moving parts. The solution came as a flash of inspiration to Colin Debenhan, an Englishman responsible for the mechanical design of the soil distribution system for the experiments in the biology instrument. At home one evening he accidentally knocked over an ashtray. As he used a brush to sweep the spilled ashes into a dustpan, he realized that a brush was the answer to the design problem on Viking. As a result the biology instrument was fitted with a small, stainless steel brush to move the soil samples in measured amounts across a steel plate to the three experiments within the instrument.

The biology package attempted to do something quite unprecedented—the automation and miniaturization of what would normally be a large biological laboratory. Despite many problems in testing the instrument, the required units were delivered on time, and two were flown as required in the Vikings to Mars. Dr. Fred Brown was the Viking biology-instrument manager at TRW Systems. It is important to understand, he said, that a biological instrument of this type is a "consumable"—it cannot be used after testing because it uses up many of its parts which cannot be replaced practically. The actual units flown could not be tested before they went to Mars; tests had been made on similar units that could never be flown. Before the Vikings were launched, many people expressed concern that such complex biological experiments might not work when they finally reached Mars, despite the tremendous amount of effort that had been expended in their design and development.

In contrast to many other federal programs (construction of new buildings, mass transit systems, health and welfare, urban renewal, defense), Viking had to reach Mars successfully without any cost overruns. This was a fantastic challenge. Whereas the other programs dealt generally with activities on which there was already much experience, e.g., building new federal offices or constructing a rapid transit system, a scientific expedition to land on another planet of the solar system was something that had never been done before; not even with a simple lander, let alone the enormously complex, fully automatic machine of Viking.

During development of Viking, inflation was high and national priorities and budgetary constraints made it essential for the project to avoid cost overruns. NASA initiated a comprehensive cost-control plan to ensure that Viking could reach Mars as planned. Some experiments were shelved or reduced in scope. In addition to eliminating Dr. Vishniac's life-seeking experiment, the program had to deprive the meteorology experiment of a movable boom and one of its sensors, a humidity detector. A rule was established that unplanned additions to the cost of any system or instrument must always be balanced by elimination of an equal cost elsewhere on Viking. The number of spacecraft to be built for development and testing was curtailed; prototypes complete in all details were not built, but only proof test models to test major subsystems. Hardware served a dual purpose of testing each subsystem and later testing the whole spacecraft.

Also the number of test teams was cut to one. This team moved from vehicle to vehicle, from the proof test vehicle to the two flight vehicles. Normally there should have been a test team for each vehicle.

Expedition to the Red Planet

The proof test orbiter was redesigned so that it could serve as a flight vehicle, since there was no money to make a spare flight vehicle. It was very fortunate that the project was able to upgrade the proof test vehicle; otherwise the mission could not have been made. During the final prelaunch activities, one of the flight vehicles had to be replaced by the upgraded test vehicle, if there were to be two launches within the launch window.

Early in 1975 the first orbiter left the Jet Propulsion Laboratory for its journey by road to Florida. It arrived at the Kennedy Space Center on February 11, six months before the launch date. Because it was ahead of schedule, tests of the orbiter with the lander could take place at Kennedy, thereby cutting costs further.

Meanwhile the landers were being built at Martin Marietta in Denver. Making the landers sturdy enough to survive the heat of sterilization caused many problems, especially in connection with the tape recorder, the batteries, and the computer. Heating a computer was unheard of; complex computers were used mainly in controlled, air-conditioned environments. The Viking lander needed an advanced computer because at times the spacecraft would have to operate on its own. Most earlier spacecraft were in continuous contact with their controllers and were commanded and operated in "real time," like driving a car or flying an airplane. But with Viking so far away, the mission controllers at Pasadena could not control the lander from the time it separated from the orbiter until it touched down on the surface of Mars. Separation would take place at a distance of 220 million miles (354 million km) from earth. Radio signals take 19.5 minutes to cover this distance one way. The Viking lander could not be flown down to the surface of Mars by mission controllers because touchdown would occur only 18 minutes after entry into the Martian atmosphere, less than the time for

radio signals to make the one-way trip to or from earth.

Even after a successful landing, the lander would be out of touch with earth for half of each Martian day because of Mars' rotation on its axis. (The Deep Space Network of tracking stations prevented similar problems from the earth's rotation.) And if something went wrong at the landing so that the spacecraft could not turn its antenna toward the earth to receive commands, it must still continue its mission automatically, sending its data to earth via the orbiter. The lander spacecraft, however, could not be commanded via the orbiter.

The lander had to be designed as both an automatic and an adaptive spacecraft. If all went well, the mission had to be capable of change to take advantage of what it found out about Mars. The lander's computer was the heart of this automatic and adaptive system. The computer required a large memory, for which a grid of fine hairlike wires was chosen. This produced a sterilization problem due to heat distortion of the wires, until these wires were protected by several platings. Because of these sterilization problems, the lander computer was not ready until close to the launch date, and an interim version had to be used in the earlier checkout testing of the lander.

A major effort was involved in developing a way to get the lander from orbit to the Martian surface. Tests began as early as 1970 to perfect a parachute and aeroshell combination. Models were dropped from aircraft. Others were fired from rockets and allowed to plunge back into the earth's atmosphere. Still others were dropped from high-altitude balloons. The system was approved in December 1972, but more work remained to be done on the three rocket engines used in final stages of descent, on stabilization of the spacecraft, and on the radar altimeter system.

Expedition to the Red Planet

4.12 When the orbiter and lander were connected at the Kennedy Space Center, in Florida, everything fitted perfectly.

In March 1975 the orbiter and lander were connected for the first time, at the Kennedy Space Center (figure 4.12). Everything fit exactly as planned—the first Viking was ready to start its extensive and thorough prelaunch testing.

After many successful tests of the spacecraft, a setback occurred during the last week of May. The building housing the orbiter was struck by lightning, and electronic components in the propulsion module of the orbiter suffered damage. This module contained propellant tanks and a rocket engine for entering Martian orbit and subsequent maneuvers. The module was replaced by the one being readied for the second launch, and the spare test module was allocated to the second Viking. The damaged module was later repaired and held for a spare. Summer thunderstorms continued to hold up work because of the routine precaution of halting all activities to protect both people and spacecraft if a storm approached within 5 miles of the Center.

June 1975 was a critical period for everyone connected with Viking. The first lander spacecraft in its aeroshell had been sealed inside its bioshield. White-coated workers pushed the spacecraft into the sterilization chamber (figure 4.13). Surrounded by swirling clouds of hot nitrogen gas, the Viking lander was baked at 235 degrees Fahrenheit (112°C) for 40 hours. After removal from the sterilization chamber, it had to be tested further. Everything seemed fine; the lander had been successfully sterilized without damage. Immediately afterward the second lander was sterilized, and it, too, came through undamaged.

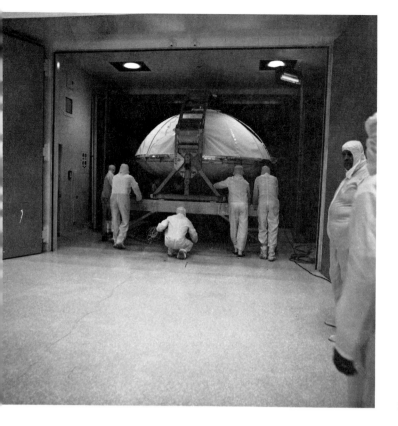

4.13 White-coated workers push the encapsulated lander spacecraft into the sterilization chamber at Kennedy Space Center to rid it of microorganisms before its flight to Mars.

One of the landers was then mated to an orbiter and became the Viking A spacecraft. NASA spacecraft are given letter designations until after launch, when they are given a number. After further testing, this complete spacecraft was placed in the protective streamlined shell of the nose fairing on July 11. The second, Viking B, spacecraft was placed in its nose fairing on July 24. Four days later, on July 28, Viking A was placed on top of the Titan/Centaur launch vehicle on the launch pad at Complex 41 (figure 4.14). Everything looked good for a launch at 1:59 P.M. on August 11.

The mission directorate gave final approval for the launching. The countdown started. But at 115 minutes before scheduled lift-off, a valve on the Titan's solid-propellant rocket booster failed to operate. The launch had to be postponed.

The faulty valve (needed to control the Titan's flight path) was removed, and the launch rescheduled for August 13. On August 12, technicians discovered that the batteries of the orbiter had been mysteriously drained of nearly all their charge. Exhaustive checks indicated that a lightning flash had produced a signal

to the spacecraft in the control wires from the command center. But there was no way of knowing whether the depleted battery had caused damage to other components of the orbiter. The spacecraft had to be taken off the launch vehicle and thoroughly inspected.

So Viking A was removed from the launch vehicle and replaced by Viking B. While technicians scurried around checking Viking A, Viking B started a countdown for a new launch date of August 20. This time the countdown proceeded without any problems. At 2:22 P.M. Pacific Daylight Time, the first of the Viking spacecraft rose into a cloudy sky on its way to Mars (figure 4.15), and became Viking 1.

The second Titan/Centaur was then moved onto the launch pad. The orbiter with the drained batteries proved to be undamaged. It was remated with its lander on August 23, and readied for launch. In the midst of precountdown testing, however, problems developed in one of the orbiter's radio receivers. The spacecraft had to be taken off the launch vehicle, and all the radio-frequency cables from the orbiter's antenna to its receiver were replaced.

A new countdown started. All went well, and the second Viking (designated Viking 2) lifted from the launch pad at 11:39 A.M. PDT on September 9 (figure 4.16). Both Vikings were at last on their way to Mars.

During the long flight of the Vikings, preparations were made on earth for conducting the expedition

Expedition to the Red Planet

4.14 Viking A is being placed onto the Titan/Centaur launch vehicle in preparation for the first launch.

when the spacecraft reached Mars. An adaptive mission strategy was developed, whereby later parts of the exploration could be changed to reflect new information discovered about Mars. The Viking mission-operations plan had to deal with four spacecraft—two orbiters and two landers. All the ground facilities, including the Deep Space Network with its tracking stations round the world, would be at their operational limits. Moreover, the difference between Martian time and earth time (a day on Mars is 1.03 times as long as a day on earth) would cause unusual work shifts because operations on Mars had to be geared to Martian time. While telemetry data could be returned from the orbiter over 24 hours each day, that from the lander came back once each day for 55 minutes only. The communications between lander and orbiter were open for only about 17 minutes each day, and the direct lander-to-earth link was open for

4.15 Viking 1 roars off its launch pad on its way to Mars.

advance. Changes were required only if problems arose, and the changes were usually made in order to return the experiment to the preplanned sequences. By contrast, scientists were not sure what to expect on the surface of Mars. As new information became available from the landers, the preplanned experiments might have to be changed in many ways.

This need to alter some experiments required means of introducing the changes into the command sequences stored within the lander's computer, without disrupting other experiments. A unique organization consisted of teams of scientists and engineers as representatives for each experiment. The team cut across institutional lines, and included people from NASA, JPL, Martin Marietta, subcontractors, and universities. The most qualified people were selected, regardless of where they were employed. This integrated team concept was tested in several simulations that trained the people first to carry out the operation plan, next to demonstrate the plan's practicality, and finally to operate the spacecraft in simulated situations. These tests simulated the arrival at Mars and the landing on the surface. They dealt with unexpected problems connected with the spacecraft and the Martian surface.

only 55 minutes each day to avoid overheating the amplifier tube in the lander's transmitter. The Martian day of 24.6 hours was called a sol to differentiate it from the terrestrial day.

In order to be on duty at about the same time each sol, controllers had to move their work hours. From an 8:00 A.M. to 4:30 P.M. shift at the beginning of the planetary mission, controllers would have to change their duty after several days to a 4:00 P.M. to midnight shift, and then two weeks later to a midnight to 8:00 A.M. shift. The cycle periodically repeated throughout the mission.

In previous space missions the way in which science instruments were to be used was clearly defined in

Meanwhile, out in space the two Vikings were being exercised. In October 1975, lander 2 was scheduled to have its batteries charged. They had been uncharged during sterilization and for the first part of the journey to prolong their operational life. But the commands to the charger were not obeyed. After extensive tests with an engineering test model of the spacecraft at Martin Marietta, engineers concluded that the battery charger itself was at fault. This was where redundancy paid dividends. A backup charger in the spacecraft was commanded to turn on, and the lander's batteries were then charged successfully.

In general, however, the long cruise through interplanetary space was uneventful, and all the scien-

Expedition to the Red Planet

4.16 The launching of Viking 2 is reflected in pools of water at the Kennedy Space Center, as it follows the first Viking toward a rendezvous with the Red Planet.

tific experiments seemed to be passing their in-flight tests.

On June 9, 1976, Viking 1 was being prepared for a critical maneuver that would take place in another 10 days—the vital insertion into orbit around Mars. If this maneuver were not successful, there could be no landing. It was at this time that engineers inspecting the many items of data telemetered from Viking to the ground noticed that helium gas pressure in the orbiter's propulsion system was rising when it should have been stable.

Helium was carried to pressurize the propellant needed for the orbiter's rocket engine. The helium was stored in a high-pressure tank and applied through a pressure-reducing valve to the propellant tank to force propellant into the combustion chamber of the rocket engine. Several pyrotechnic valves controlled the flow of gas from the helium tank to the pressure regulator; each could be fired only once. Their purpose was to isolate the helium tank from the regulator during the long interplanetary flight, and then to open the gas line when the spacecraft approached Mars. The line was now open in prepara-

tion for the maneuver into orbit around Mars. Two pyrotechnic valves remained unfired, one to close the helium line and another to open it again.

The slow rise in helium gas pressure was thought to be caused by a small piece of material lodged in the seal of the pressure regulator. It was not an emergency but it required attention. One way would have been to fire the remaining shutoff valve and close the helium line to the pressure regulator. But then there would remain only one valve to open it again. If this valve should fail to work, Viking itself would fail. Although these valves were known to be extemely reliable, the project manager, Jim Martin, who had responsibility for the overall Viking project, decided it was not prudent to take the risk. Instead he asked that a new approach maneuver to Mars be planned and executed.

For this new approach, the course correction maneuver was delayed from June 9 until the following day.

Expedition to the Red Planet

The delay meant that the rocket engine of the orbiter would need a longer period of thrust, and this would use more propellant and more helium gas, thereby slowing the rise in helium pressure. After the rocket burn, however, the helium pressure continued to rise. The rocket engine was ignited again on June 15, using up enough helium to keep the pressure from reaching a problematic level before the maneuver to place the Viking into orbit. During that maneuver, so much propellant and helium gas would be consumed that the faulty pressure regulator could not cause any further problems.

The two thrusts of the rocket engine, on June 10 and June 15, had slowed Viking down more than originally planned for the approach to Mars. Speeding it up again to meet the arrival time originally scheduled would have wasted propellant. Instead Martin let Viking arrive at Mars later than planned but had it enter into orbit just one orbit period later than originally intended and also began its site certification operations one orbit earlier than originally intended. Thus the timetable for certifying the landing sites from orbit remained almost unchanged. The adaptability of Viking was well used.

Chapter 5

Where to Land—The Blandlands

Following the orbital success of Mariner 9, the view of Mars had changed completely. The high concentrations of volcanoes near the Tharsis ridge intrigued volcanologists. The great rift valley of Valles Marineris and its many associated canyons showed erosion on a massive scale. Many channels on the planet's surface were difficult to explain except as the result of great floods cutting across Mars in the past. The polar regions appeared to be layered terrain and resembled rosettes on the north and south of the planet. A wide belt around the northern latitudes appeared much lower than the rest of Mars and might be the equivalent of a terrestrial ocean basin. In the southern hemisphere there were several large circular plains, one of which, called Hellas, had a rather featureless floor in the spacecraft's pictures. Was it a huge dust bowl or a smooth lava surface? Were these great basins the remains of impact basins similar to those on the moon and Mercury?

Many areas of Mars looked as though they might have resulted from the collapse of underlying material that tumbled the surface into a mass of blocks. Geologists wondered whether this chaotic terrain was caused by melting of underground ice which might have provided the water that cut the Martian channels.

The geology of Mars intrigued geologists and planetologists alike because of its great variety and the many unusual features that defied clearcut or easy explanation. The variety was fascinating, but the Viking mission had only two lander spacecraft to explore it with. A landing in the polar regions or on the floor of a big canyon, the caldera of a huge volcano, or the bed of one of the wide channels would have been of great geological interest. In terms of biology, a landing should be in a low-lying area where water might exist today.

The spacecraft itself presented limitations. The landers should go to low-lying places on Mars to use the maximum density of the atmosphere for braking purposes. Also, the Viking lander was a very squat machine and needed a relatively smooth surface for safe landing. It could not be sent into a boulder field or into soft material that might engulf it. It could not be sent close to mountains or canyon walls because possible errors in aiming were quite large when related to distances on the surface of Mars. The ellipse within which the Viking was originally expected to land was 370 miles long by 50 miles wide (600 by 80 km), to allow for uncertainties in navigation, in the atmosphere of Mars, in the shape of Mars, and in elevations on the Martian surface. This is equivalent to targeting the landing for Central Park in the center of Manhattan, and hoping the spacecraft will land somewhere between Baltimore and Boston, or between Paterson, New Jersey and Long Island. The difficult task was to find this large an area on which the terrain did not vary enough to present a hazard to the spacecraft.

As the mission progressed, navigators were able to narrow the landing uncertainties. When Viking was successfully in orbit, they established a 99 percent probability of being able to land within an ellipse 143 miles long and 62 miles wide (230 by 100 km), and a 50 percent probability of being able to land within an ellipse 58 miles long and 25 miles wide (94 by 40 km).

The search for suitable landing sites started early in the program. A landing-site working group, chaired by Thomas Young of NASA's Langley Research Center (who later became mission director, responsible for the mission aspects of the Viking expedition), involved project personnel and all the scientists who had experiments on the spacecraft. The group first met in August 1972. Concepts of Mars were changing rapidly at the time they were selecting the landing sites for Viking. These changes resulted from analysis of the Mariner 9 data. It was determined that the spacecraft could be aimed to land at sites anywhere on Mars between 25 degrees south and 75 degrees north. Initially the group concentrated on the equatorial belt because they had most information about that region. But more data from Mariner 9 caused the group to expand its survey. The polar region from 60 to 75 degrees north was interesting from the biological standpoint because water might be available at the boundaries of the polar cap as it melted.

Ground rules were established for the choice of landing sites. Because the equatorial regions were known best, the project management decided that the first

Viking should land between 25 degrees north and 25 degrees south of the equator. The second Viking might then be targeted to land anywhere in the region between 25 and 75 degrees north. But if the first landing site did not prove successful, the second Viking would also be sent to an equatorial site.

Scientific criteria for selecting the landing sites suited the biological emphasis of the mission: to find sites that were in some way correlated with water, if possible liquid water. Also the landing site should have a low elevation to keep the atmospheric pressure as high as possible and thus make it more likely that liquid water would be present.

The third criterion was to land close to or on a "river" delta, such as the mouth of one of the channels discovered by Mariner. If these channels had been caused by water flowing on Mars, their mouths would be rich areas for scientific experiments.

Other scientific criteria were that the two landing sites should differ geologically; both should be unobstructed by mountains to permit good meteorological experiments; and finally, to help the seismology experiment, the sites should not be too far from each other.

Ideally a Viking landing site should have slopes of less than 19 degrees, it should be free of rocks, and the surface material should be able to support the impact of the landing. The surface should, however, be of material that could be sampled—it should not be flat rock. The site should also be relatively wind free.

The geology and terrain at 22 potential sites were studied extensively, using the Mariner pictures. Eleven of the sites were within the equatorial band, seven in middle latitudes, and four in the north polar region.

Where to Land—The Blandlands

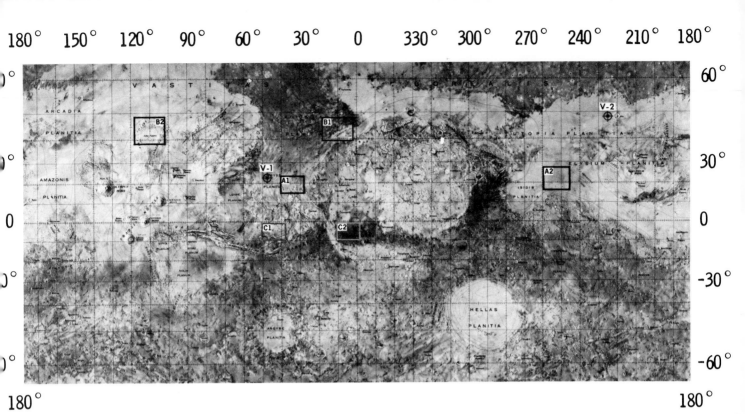

180° 150° 120° 90° 60° 30° 0 330° 300° 270° 240° 210° 180°

60°

30°

0

-30°

-60°

180°   180°

5.1 Landing sites on Mars were selected for Viking several years before the launching of the spacecraft. These consisted of primary and secondary sites for the first spacecraft (A-1 and A-2) and for the second spacecraft (B-1 and B-2). In addition, two alternate sites (C-1 and C-2) were picked to use if observations from orbit showed that the sites in higher northern latitudes were not suitable for landings. The locations of the actual landings (V-1 and V-2) are also shown on this map.

On May 7, 1973, John Naugle, Associate Administrator of NASA, announced at a press conference that an area of the Chryse Planitia had been picked for the first Viking (mission A) and an area known as Cydonia for the second spacecraft (mission B). Two backup sites (figure 5.1) were also chosen.

Although the nomenclature for Martian surface features is gradually being updated as a result of discoveries by spacecraft, at the time of Viking it still consisted of an illogical mix of classical names (which bore no real reassemblance to the actual geological features) and newer terminology based on geology. Early astronomers had called the dark areas maria (seas) and the light areas deserts, and classical terms were used, such as lacus (lake), sinus (bay), and fons (spring). Names were also drawn from mythology to refer to complete "districts" of Mars, such as Elysium, Hellas, Arcadia, and Chryse. As the true nature of some of these features became known, they were given more representative names, e.g., Chryse became Chryse Planitia (plain), Coprates Chasma (chasm, canyon), Ares Vallis (valley), Elysium Fossae (riverbed or watercourse), and Pavonis Mons (mountain).

The prime site (which NASA called A-1) for the first spacecraft was located in Chryse Planitia at 34 degrees west longitude and 19.5 degrees north latitude. The backup site (A-2) for this spacecraft was in Tritonis Lacus, at 252 degrees west longitude and 20 degrees north latitude.

Dr. Gerald Soffen, the project scientist, commented that "we never had any trouble with mission 'A.' We all recognized it must go to a safe place, the safest place we could find, and the scientifically most interesting place. We decided very early it would be in the northern hemisphere. We decided very early it would go to a very deep [low-elevation] area." Chryse was this type of site. The name means "land of gold" and refers to a golden land in the far east. In mythology,

Where to Land—The Blandlands

Chryse was a priest of Apollo whose daughter was seized in the battle of Troy and given to Agamemnon, only to be returned after Apollo struck the Greek camp with a plague.

Mariner 9 pictures showed that the southern half of the Chryse area consisted of plains broken up by many channels. Much material seemed to have been swept northward along well-defined winding valleys to a low area of only slight relief on which the landing ellipse for the first Viking fitted nicely. Scientists believed from the available evidence that most of the surface at the site was partially covered by wind-transported dust deposits interspersed with material washed from the great canyons of Valles Marineris. There might also be sand dunes in the area. The site is a region where water might have flowed in large amounts in the past.

The Tritonis Lacus backup site for the first landing was in a similar geological area, to the east of the Syrtis Major and not far from the volcanic plateau known as Elysium Planitia.

The prime site (B-1) for the second spacecraft was chosen where water might be available today to support Martian biology. Near 40 degrees north latitude, water might be liquid for part of the Martian year. In a band stretching around the planet at about that latitude there were large broad valleys in which water might be concentrated. Measurements of atmospheric water vapor over the valleys seemed to bear out this possibility. The site selected in Cydonia was a flat stretch of the northern basin plains. Cydonia is the name of a town in Crete which, in turn, was named after Kydon, the son of the greatest king of Crete, Minos.

The Cydonia landing site consisted of smooth and mottled rolling plains, possibly flows of basalt (volcanic rock) covered by wind-blown debris, volcanic dust, and water-borne sediments. There might be some volcanic cones. The site was located on the eastern side of the Mare Acidalium, where regions of plains of the Martian northern lowlands meet the higher equatorial plateaus and hills. The Cydonia site was centered at 10 degrees west longitude and 44.3 degrees north latitude. The backup site (B-2) for the second spacecraft was selected in the region of Alba Patera, a volcanic, bowl-shaped plateau in the same northern band of Mars, at 110 degrees west longitude.

When the Vikings were launched, no further information had become available to warrant changing these preselected sites. However, two C sites (shown in figure 5.1) had also been selected in the equatorial zone at 5 and 45 degrees west longitude. They were considered the best sites to land a spacecraft on Mars if surveys from orbit showed that the A and B sites were hazardous. The C sites were not good, however, in terms of water and biology because their elevation was too great for water to exist today and their geology did not suggest any major flow of water there in the past.

As the two Viking spacecraft approached Mars, I talked with Harold Masursky, of the U.S. Geological Survey in Flagstaff, Arizona, who headed a team of scientists responsible for certifying the sites before a landing was attempted. Dr. Masursky explained that site certification relied on several tools—photographs and infrared scans from orbit and radar observations from earth, plus a lot of geological intuition. Earth-based radar had inspected the Chryse area in 1967. These early surveys indicated that the Chryse site and its backup in Tritonus Lacus had quite different radar characteristics.

More powerful radar probing of the Viking landing sites had to wait until just before the first Viking spacecraft arrived at Mars, because the position of

Where to Land—The Blandlands

Mars on its orbit and the inclination of the planet's axis did not allow the landing sites to face directly toward earth until then. The site must face earth for a radar echo to be received. But by that time the distance to Mars was so great (228 million miles, 380 million km) that the signal reflected from Mars was almost drowned by interference from natural radio sources. This was, nevertheless, the only opportunity to prove the sites with radar before attempting a landing.

From May 29 through June 12, 1976, the 210-foot (64-m) antenna of the Deep Space Network's Goldstone tracking station in the California desert periodically bounced radar echoes off the region of Chryse, and from May 11 through June 15 it also covered the backup region. Later the 1,000-foot (300-m) diameter steel-mesh antenna that is suspended in a great limestone bowl at Arecibo, Puerto Rico, also examined the sites. The Goldstone antenna used a frequency of 8,400 megahertz and the Arecibo antenna used 2,400 megahertz.

The surface of Mars, said Dr. Masursky, was known to be about five times as rough as that of the moon. This difference applied to elevations of major landforms on Mars and also to small-scale surface features, as revealed by the radar reflections. However, what was seen in the Mariner pictures taken from orbit and what the radar results implied did not appear to match in any clear way. Some areas that looked smooth optically gave rough radar echoes; others gave smooth echoes.

A check of terrestrial landforms explained the anomaly. Radar reflections returned to high-flying aircraft from dry riverbeds in Arizona, and from sand dunes in California, matched optically with and produced radar signatures similar to features on Mars. Martian plains that produced extreme radar scattering were believed to possess steep slopes that were too small to be visible on the photographs. Some of these rough plains were thought to be sand dunes, because their radar signature matched that of terrestrial dune fields. Extreme radar scattering in the Martian highlands was thought to be caused by tiny channels like the dry washes of Arizona.

Site certification became a dramatic interplay between safety and science—the structure and density of the Martian surface at the site (which would affect the ability of Viking to land on it and provide an observation platform after the landing) and the potential of the site to provide scientific information about Mars and its history.

No landing site could be found on Mars that appeared completely free of hazards to the spacecraft. The task during site certification was to compare all the data and try to reduce the hazards as much as possible.

As Mars traveled along its orbit, the point from which radar signals were reflected, known as the radar bounce point, moved north across the planet's surface at the rate of 0.25 degrees per day. Since Mars was so far away, multiple traces were desired across each site to improve the data. Radar could not cover any of the sites for the second landing, however, because they were too far north.

The prime task was to certify a site in Chryse (figure 5.2) for the first lander. The site certification team had rather mixed feelings about the safety of the Chryse landing site. Dr. Masursky commented that the Mariner pictures just did not provide sufficient detail. He also talked about the geological knowledge that might be obtained by sampling the surface material of Mars. If there were sand dunes at the landing sites, the material might have originated in the highlands and been swept down by water. If there was sandy material on the great lava plain of the second

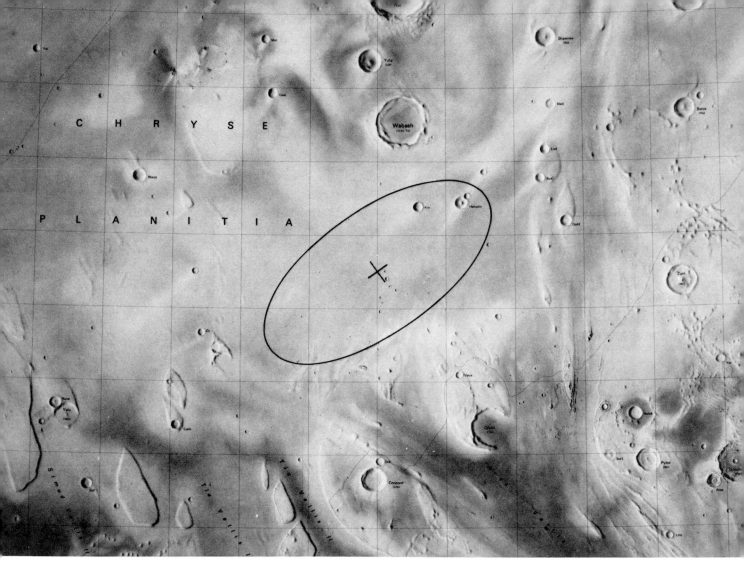

5.2 A typical landing ellipse is shown superimposed on a U.S. Geological Survey map of Chryse Planitia (the primary site for the landing of Viking 1). The map was drawn from pictures obtained by Mariner 9.

landing site, this might be from basaltic rock. However, if the surface of Mars was covered with silt, the same material might occur all over the planet; it might have been carried high into the atmosphere, thoroughly mixed, and deposited everywhere. That would be the worst case from the geological standpoint. It would still be of interest, however, because it would provide information about an average chemistry of Mars. Ideally, the scientists wanted to be able to analyze samples from two different sites that would provide information about both highland material and plains material.

The preselected Chryse site was thought to be a good geological site and a safe site. If the landing were moved closer to the highlands, Viking might land on boulders; if the landing were moved farther down-

stream, it might set down on sand dunes. But generally the team believed, said Dr. Masursky, that this was a very flat region, like a giant apron; it had slopes, but gentle ones. They would check it with radar, with infrared, and with orbital pictures. They hoped to land close enough to one of the craters in the area to sample material ejected from it by the impact which dug the crater.

On the matter of choosing between B and C sites for the second landing, Dr. Masursky said that the main criterion was the availability of water on Mars. The

Where to Land—The Blandlands

highest amount of water vapor found in the Martian atmosphere by Mariner 9 was in the northern lava plains. These plains could be smooth territory similar to the Oceanus Procellarum on the moon. But many things in the Mariner pictures of this area where they wanted to land, such as ridges and lava flows, made it look more like the rougher Imbrium plain on the moon.

Also, the northern plains have what are called pedestal craters. The material ejected from a crater at the time of its formation is called an ejecta blanket. These blankets of debris surrounding a crater protected the Martian surface beneath it. When wind eroded the surrounding territory, the ejecta blankets were not eroded. Today the craters are left standing high above the rest of the terrain. The many small pedestal craters in the northern plains may make them difficult sites for a landing, explained Dr. Masursky.

The backup site for lander 1 was in an entirely different type of area from the prime site. Alba Patera was thought to be a volcanic plateau, but very different from the big volcanoes of the Tharsis region. It looked similar to what are known as ring dyke complexes on earth, in which the center of the volcanic caldera has sagged to form a shallow bowl surrounded by rings of lava dykes. The whole plateau of Alba Patera was surrounded by a complex fracture system; some fractures were concentric and other radial to the caldera. It was also an area of persistent white clouds, which gave rise to its name, "white bowl."

The C sites near the equator were inspected with radar and seemed to be quite smooth. The prime C site (i.e., first choice) was north of Ganges Chasma in a shallow valley with sides that sloped about 6 degrees. The site was in low highlands at one of the few places on Mars where the highland rocks were at a low enough elevation to permit a Viking landing on this type of ancient Martian crust. The backup C site, also in the equatorial zone, was in low highlands of Sinus Meridiani, and its radar reflectivity made it look very smooth. The decision between landing at B or C was basically a question of water, which might be present at the B sites but was unlikely at the C sites.

In summary, said Masursky, each of the Martian sites had some very attractive scientific possibilities, and each had its strong points for and against safety. There were different kinds of hazards, but no place had been found that was free of hazard to a landing.

Viking 1 was still committed to land at Chryse when the spacecraft entered orbit around Mars on June 19, 1976. Then the process of site certification became intense. During the third revolution about Mars, on June 22, the orbiter concentrated on photographing the Chryse area in detail with 58 overlapping, high-resolution pictures. The site certification team was disappointed. Chryse did not look like such a good place to land after all.

The drama began to build at conferences on the third floor of the Viking Building at the Jet Propulsion Laboratory, where the orbiter imaging team had their offices. By the fifth orbit of Mars, more than 160 pictures had been taken and returned to earth. Members of the team had pieced the pictures together like a huge jigsaw puzzle, carefully matching the overlaps (figure 5.3).

By June 23, project officials were expressing surprise at the features revealed in the pictures. The plain of Chryse was not the bland, flat area hoped for. The terrain was rugged, cratered, with deep channels that looked like dry streambeds. Even though by this time the space navigators were confident they could land on Mars within an ellipse that was 75 miles long by

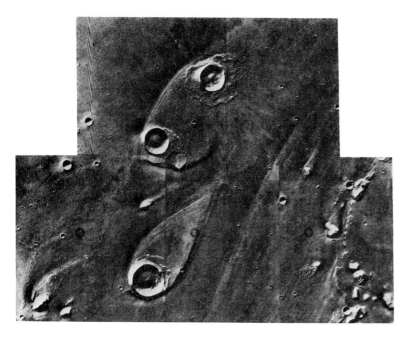

5.3 *At top,* Dr. Michael Carr (leader of the orbiter imaging team) points to hazards at the preselected landing site for Viking 1, and to the flow lines around crater Gold on Chryse Planitia. The big crater with eroded walls is Wabash (see also figure 5.2). *At bottom,* that area is shown in greater detail. Braided channels record water flowing sometime during the past. Fine grooves and hollows can be seen on the upstream side of obstacles to the flow.

31 miles wide (120 by 50 km), compared with the several-hundred-mile-long ellipse considered likely at the start of the mission, the southeastern plain of Chryse, selected before the mission, did not have any safe area in which to place this smaller ellipse.

The pictures appeared to confirm that immense floods had rushed through the Chryse basin; meandering channels and deeply grooved narrow sections suggested liquid flow on a grand scale (figure 5.4). The likely liquid was water. Deep grooving at the side of some channels could only have been made by fast-flowing streams, by a turbulent flow of water at high speed.

In subsequent days, as more areas were photographed from orbit, the site certification team became more optimistic. Dr. Michael Carr, leader of the orbiter imaging team, commented that the surface did not look as bad elsewhere as it had on the first pictures. Although the preselected landing site was interesting geologically, it was not acceptable in terms of safety. Jim Martin, the project manager, had to decide soon whether or not Viking could make its bicentennial landing on July 4.

On Sunday, June 27, I chaired a one-day symposium designed to stimulate public interest in the space program, and particularly in the Viking mission and what it implied for the future. It was a meeting of the Southern California branch of the British Interplanetary Society, held at La Cañada High School, a short distance from the Jet Propulsion Laboratory. I introduced speakers such as Gerry Soffen, Viking's chief scientist, Ray Bradbury, a science fiction author, and Jim Martin. Other speakers included Dr. William Pickering, the former director of JPL, Dr. Robert Forward of Hughes Research Laboratories, an advocate of interstellar flight, and Dr. Bernard Oliver of Hewlett Packard, an advocate of searching by radio for extraterrestrial civilizations.

Where to Land—The Blandlands

The first speaker was Jim Martin. Before he discussed the Viking program, he announced that the July 4 landing on Mars must be delayed. He had just come from a press conference at which he had made the same statement to the mass media representatives. But the hundreds of people at the symposium were the first of the general public to hear the news. Mr. Martin said he felt that the terrain in the preselected landing area was too hazardous and he wanted to inspect an area northwest of it.

Three plans were later evolved for the delayed landing. The earliest possible time for landing was now July 9. Other alternatives would delay the landing about two weeks. By noon (PDT) on July 1, Jim Martin had to decide which of two alternate sites would be kept open as possibilities—a site about 185 miles (300 km) northwest of the A-1 site, or the backup site (A-2) in Tritonis Lacus. Landing at either site would require firing the spacecraft's rocket engine to change the orbit slightly. The Tritonis Lacus site had been surveyed by radar, and although the site looked rough on the Mariner 9 pictures, it appeared smooth on the radar.

On July 1, Dr. James Fletcher, Administrator of NASA, announced at a press conference at JPL that the A-1 northwest site had been selected as the second possibility. Jim Martin said that the expedition now aimed for a July 17 landing. He described the site team's problems with understanding the things they saw on Mars. They had struggled with the question of what was happening on the planet. Dr. Robert Hargraves had suggested, "If you don't like the area where the deposits are leaving, why not go to the area where they are settling." So the team had looked where the river went through Chryse, toward the northwest, and started taking pictures of that area. "We are pleased with what we see and find it more understandable," said Jim Martin, adding that he believed the data could be extrapolated from the 460-

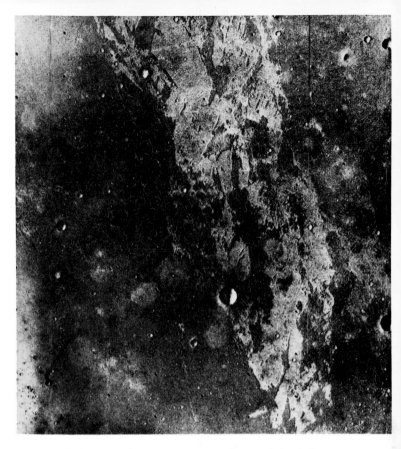

5.4 High-resolution enlargement of part of the A-1 preselected site within the proposed landing ellipse. These scablands with their steep cliffs, pockmarked everywhere with small craters, would have presented a serious hazard to the Viking lander.

foot (140-m) resolution of the orbiter's cameras down to the lander scale of 6.5 feet (2 m). "I believe we have every opportunity for a successful landing."

Dr. Masursky estimated a 1 to 2 percent probability of landing on a hazard at the northwest site, compared with 9 to 11 percent at the preselected site.

The "northwest territory" (A-1 northwest), explained Michael Carr at a briefing, appeared to have fewer small craters than the primary site. However, there were surface cracks, escarpments (steep slopes or low cliffs), knob features, and dunes. The relief was mostly gentler than at the primary site (figure 5.5).

A decision to land at one of the alternate sites had to be made by Monday July 5, because later that day the spacecraft would need to make the appropriate engine burn to trim its orbit. It was still possible,

Where to Land—The Blandlands

however, that a landing ellipse might be found in the southwest section of the prime landing site.

Two problems faced the Viking team. First, the spacecraft had only a limited amount of propellant for maneuvering, and changing to different sites to examine them closely and then adjusting the orbit for a landing would use up propellant. Project scientists wanted to save some propellant for manuevering the orbiter after the lander was down on the surface, so that the orbiter experiments could obtain maximum benefits by covering more of the planet. Also, Viking 2 was only 5.5 million miles (8.9 million km) from Mars and was approaching rapidly. It would have to be placed in orbit on August 7, requiring full concentration for a while. So if Viking 1 could not land before July 25, it would have to wait until after the orbit insertion of Viking 2. Time was crucial because solar conjunction (with Mars and the earth on opposite sides of the sun) would take place in mid-November. The nominal missions of both landers on the surface had to be completed by that date because there was no guarantee that the landers could be recommanded into activity after a dormant period around conjunction.

Jim Martin was still optimistic about a safe landing: "We are confident we will find a safe place to land. Like Columbus, we don't rush into the first beach we see, but we look for a safe harbor." He wanted radar data before committing the spacecraft to burn its rocket engine and use about one-third of its propellant reserve to alter the orbit for a landing on a different site.

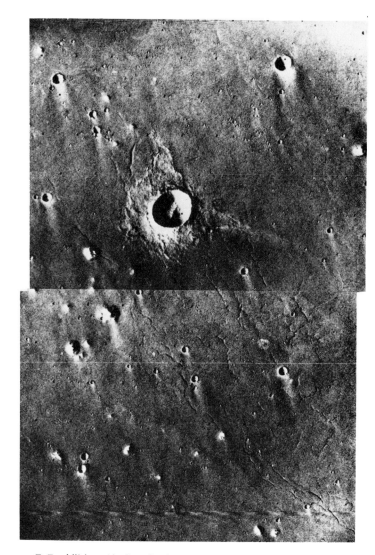

5.5 Viking 1's first look at an alternate site was obtained June 27. This picture covers an area of about 31 by 46 miles (50 by 75 km) that is some 185 miles (300 km) northwest of the primary site. An impact crater and its ejecta blanket were the main features of the area. There also were many small craters with bright wind tails. The area was thought to be covered by fractured lava fields through which knobs of older rocks projected.

5.6 Many meetings were still taking place between NASA and project officials and the site certification scientists; *left to right* (seated at the table) are Drs. Bradford Smith, Harold Masursky, and Michael Carr, talking to Gentry Lee (with back to camera).

An Arecibo radar beam bounced off the northwest area on July 2 but did not get data on the site itself, only to the west of it. On July 3 and 4, good data were received. The site appeared twice as rough as the average surface of Mars, but there was smooth surface on either side of it.

The radar data were accepted as valid because in 1967 another radar (Haystack) had looked at the same area and obtained similar results. Mars did not appear to have changed in this area over the 9 years between the two observations. Jim Martin said, "The radar is telling us something, but we are not sure what. There are so few radar experts able to interpret these radar results."

The question became whether to land at the northwest site on July 17 or to move farther west. Officials and scientists of NASA and the project held many meetings discussing all aspects of the problem (figure 5.6). On July 7 they decided to examine the area west of the northwest site.

Finally, after 25 days in orbit, during which many possible sites were surveyed in the Chryse Planitia and elsewhere, Viking 1 found a safe haven on Mars for its lander. The landing was planned for 5:12 A.M. PDT on July 20, at 22.4 degrees north and 47.5 degrees west (figure 5.7), which was about 150 miles (240 km) due west of the A-1 northwest site and still within the Chryse basin.

The night of July 19–20, 1976 was unforgettable to anyone present at the Jet Propulsion Laboratory for the first landing of an American spacecraft on Mars. It was also the seventh anniversary of the first manned landing on the moon. I had spent a busy evening addressing a group of educators invited by NASA for a prelaunch conference. The educators, who had traveled from all over the United States, had listened to project personnel talk about the mechanics

and the science of the Viking mission to Mars. I had given a keynote speech at the banquet, on what it all meant in the context of mankind's expansion from the cradle of the earth toward an interplanetary society. After the banquet, everyone wanted to snatch a few hours of sleep before the landing sequence began.

When we returned to JPL, I went to the Von Karman auditorium as a member of the news media, reporting the landing for the British journal *New Scientist*. My wife joined the educators at the Holiday Inn and went with them by bus to the JPL cafeteria where the landing events would be piped in and displayed on video screens. In other halls in the JPL complex, many people were preparing for the landing. The Laboratory was also the gathering place of hundreds of scientists, engineers, and VIP guests. There were stars from the *Star Trek* series, famous scientists and science fiction writers. The press corps was there in force, with reporters from the United States and overseas, together with commentators and other television personalities.

At 1:15 A.M., when the orbiter was not far from the highest point of its elliptical orbit around Mars, the lander separated from it and slowly began the long journey down to the planet's surface. The lander gradually accelerated, and a little more than three hours later, still protected by the aeroshell, it plunged at high speed into the Martian atmosphere. This took place at about 152 miles (245 km) altitude. The aeroshell slowed the meteoric plunge through the denser atmosphere, as its surface heated and burned off layer by layer. In 7 minutes the lander slowed enough to deploy its parachute at about 19,400 feet (5,900 m) above the surface and to jettison the aeroshell.

On-board radar was now sensing the height of the lander above mountains at the edge of the Chryse

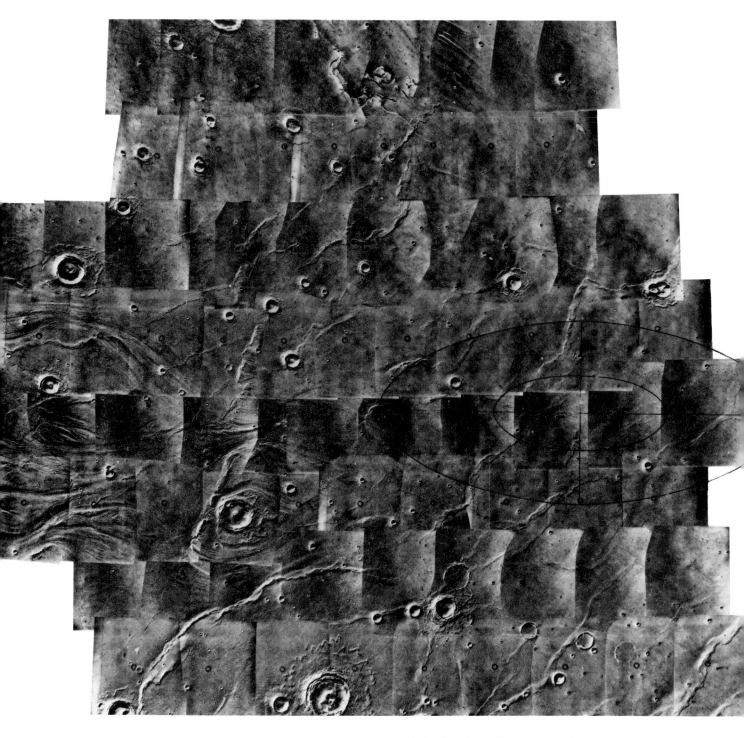

5.7 A safe landing site was finally found for Viking 1, toward the western mountains of Chryse basin. The center of the landing ellipse shown on this mosaic is 22.4 degrees north latitude, 47.5 degrees west longitude. This was 460 miles (740 km) northwest of the original preselected site for a July 4 landing, and 150 miles (240 km) west of the site selected for a July 17 landing. Landing at this new site was scheduled for July 20. Radar scans from earth indicated that the surface was smooth in this area.

5.8 Artist Don Davis captures the fantastic scene of the first spacecraft from earth to land successfully on the Red Planet Mars. An instant before touchdown, Viking lander 1's three rocket engines bring the complex machine gently toward the desertlike surface. In the distance the jettisoned aeroshell falls to the surface on the parachute. The time was 5:11 A.M., PDT, July 20, 1976.

plain and feeding the information into its computer. We on earth did not know for sure that all this was happening because the radio signals took 18 minutes to reach the earth (in fact, by the time we knew that the lander had entered the Martian atmosphere it had already reached the surface). Several events took place automatically in quick succession; the three terminal-descent rocket engines ignited, the lander's legs were deployed, and the parachute was jettisoned. Lander 1 proceeded toward the surface, slowed further by its rocket engines (figure 5.8).

As the squat landing capsule of Viking 1 streaked through the Martian atmosphere toward the surface of Mars, top officials of NASA had crowded into Jim Martin's small office in the Space Flight Operations Facility at JPL, anxiously looking for signs on the TV display screens that would tell them that the spacecraft had successfully landed (figure 5.9). In front of Martin there were rows of buttons by which he could inspect data flowing from the spacecraft as they were processed by the complex computers at the Labora-

Where to Land—The Blandlands

5.9 Dr. James C. Fletcher, NASA Administrator (*left*), and Dr. John E. Naugle, Associate Administrator, watch with Jim Martin for the computer-generated displays that will tell of Viking 1's safe landing on Mars.

tory. He looked for signs of a quick series of automatic actions that should take place within the lander in the 12 seconds following a successful touchdown.

The first indication that his years of dedication to this expedition had borne fruit came when the communications net operator announced that the rate of data transmission from lander to orbiter had changed to 16,000 bits per second. This showed that the computer aboard the lander was working and had sensed a safe touchdown. Almost at the same time, one of the items displayed on Jim Martin's viewing screen changed to "on." This told him that a signal had passed from sensors on one of the footpads to the lander's computer, telling it that the soil of Mars had been touched and the computer should start its landed operations. Other displays indicated that the heaters on the rocket engines of the lander had been switched off, and that less power was being drawn from the lander's power supply, now that it was on the surface of Mars.

Touchdown took place at 5:11 A.M. PDT, earth received time, i.e., the time the touchdown signals arrived at earth, when we knew that Viking had landed successfully on Mars, and when cheers roared

through the JPL complex. At last Jim Martin could relax. After his many years of work, the spacecraft was on Mars and operating successfully. People crowded round to congratulate him; now the exploration of Mars could really begin.

With Viking 1 down and operating on the surface, the mission then concentrated on surface operations, which are described in the next chapter. Meanwhile Viking 2 was approaching closer and closer to Mars and it, too, needed a suitable landing site. The surveys made with Viking 1 had not proved encouraging as far as Cydonia was concerned. The area was pockmarked with pedestal craters, and flat areas between these craters were broken by extensive fractures of unknown origin. On August 18, Jim Martin rejected the Cydonia region as a landing site. In general the region seemed to have ancient lava flows overlain by younger flows and later eroded to reveal ancient rough cratered terrain. The many pedestal craters would be hazardous to any landing there.

Other candidate sites were the backup site in Alba Patera (B-2), an area in Arcadia, and a region called Utopia. These sites were in the high northern latitude band at about 44 degrees north, which had been picked because liquid water might be present at the surface.

Alba Patera (figure 5.10) proved unsuitable. "We have looked at a fair number of pictures now of the B-2 area, primarily in the Alba Patera region, and that site, to me, also does not look acceptable," commented Jim Martin a few days later. The area had a complex system of faults and proved to be extremely rugged and hazardous on close inspection.

Viking 2 had to be placed in an orbit from which it could make a landing between 40 and 50 degrees north latitude. Once the spacecraft was in orbit, of course, it was not feasible to change the latitude of

Where to Land—The Blandlands

the landing site, although the longitude could be placed anywhere around the planet.

In addition to photographs from orbit, site certification for Viking 2 relied a great deal on thermal and water-vapor mapping, since the question of water availability was of paramount importance for the choice of site. The infrared thermal mapper on the first orbiter (Viking 1), normally used to provide information about the surface temperature of Mars, was being used to supplement radar data because its measurements could be interpreted in terms of surface roughness. The smaller the particles, the warmer the Martian surface would become during the daytime. By contrast, rocks absorb and radiate heat more slowly and would lag behind the daily sunlight cycle, taking longer to reach their highest temperature and longer to cool down after sunset.

The atmospheric water detector was used to ascertain where water concentrated at high northern latitudes. The scientists found that water vapor concentrated in two areas where there were low-lying valleys running approximately north–south. The area of Utopia, a candidate landing site, was one of these water-rich areas.

Viking 2 began photoreconnaissance for its lander on August 11 by photographing an area of Arcadia, northwest of Olympus Mons. At this point Viking 2 had three possible landing dates (September 3, 4, or 7), depending on the choice of site among three candidates. There were still no new pictures of Utopia, only three frames taken with the Mariner spacecraft in 1971. But if a suitable landing site could be found in Utopia (now called B-3), a landing could be made there on September 3.

Tom Young, the mission director, noted that the search for a suitable landing site for Viking 2 had covered seven times the area of the Viking 1 recon-

naissance. This enormous search showed that no areas of the Cydonia site could accommodate the landing ellipse safely. There were some possibilities in the Alba Patera region, but they were not ideal. As the mapping of Utopia proceeded (figure 5.11), several candidate areas for landing ellipses were identified.

Tom Young pointed out that finding landing sites on Mars was a difficult job. Much more time was spent in choosing a site for Viking 2 than for Viking 1. "We actually first photographed the B-1 [Cydonia] area on rev[olution] 9 of orbiter's operations, i.e., about June 27. When we put all the information together . . . the B-3 [Utopia] area, the one that's near the crater Mei, appears to be muted by a covering of material . . . [which] covers up most of the hazards that would have been there prior to the depositing of the material."

Upstairs in the Viking Building at JPL is a room that served as the meeting room for operational problem-solving (figure 5.12). Later it became a Viking museum. On the week of August 22, the room was crowded with scientists and project personnel seated at the large U-shaped table. James Martin was at the head, Tom Young, Gentry Lee, and Mike Carr close by. Several reporters had been invited to attend. The subject: where to land Viking 2 on Mars. The decision had to be made, it could no longer be put off.

Two areas were still open for discussion—Arcadia and Utopia. Scientists who had been studying the orbiters' pictures of both areas made their presentations. Dr. Barney Farmer talked about the major drainage channels on Mars. One leads northward through the Chryse area, another from Elysium, and yet another northward through Amazonis and Arcadia. They meet and continue toward the north pole. Water-vapor measurements around the planet along 45 degrees north latitude showed concentrations over

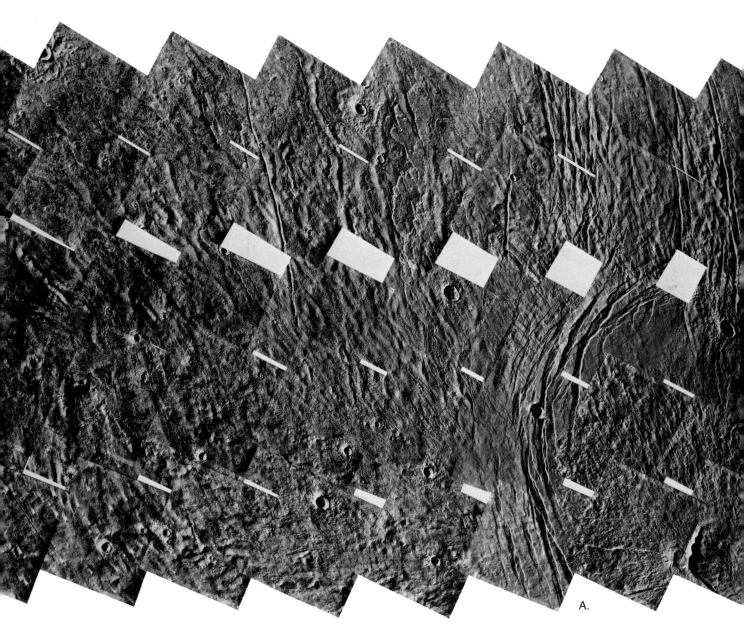

A.

5.10 Cydonia had already proved unsuitable as a landing site for Viking 2, so the search turned to the B-2 site, Alba Patera. Shown in this series of three pictures, Alba Patera turned out to be a very rugged volcanic plateau broken by many faults. Picture *A* is a general mosaic of the area west of the caldera; *B* shows an enlarged frame of the crater on the concentric fault system; *C* shows details on the western flanks farther from the caldera. Small channels are offset by the faults, showing that they are older than the faults. There are also several different ages of fault systems. Some craters are on top of faults, others are cut by faults. The whole area was much too hazardous for Viking to land, though extremely interesting from the geological standpoint.

B.

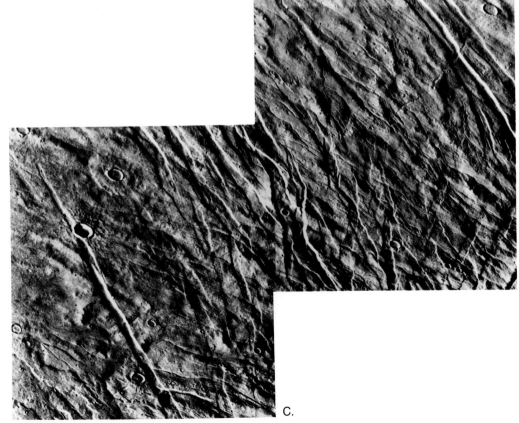

C.

such lower elevations. The Alba Patera area was quite dry, and Utopia had only about half the water vapor of Arcadia and Cydonia.

Dr. Masursky stated that the whole area of the landing ellipse at Utopia was blanketed with dune material. This material seemed obvious on the orbiter photographs, especially in the crater Mei, but thinning out and disappearing at some distance from the crater, south of the southernmost end of the proposed landing ellipse. Crater counts were consistent with a mantle in this region that disappeared to the south and to the west. Some infrared observations were available, but they were not definitive.

Unfortunately the pictures of Arcadia were marred by clouds, so Masursky's team could not be certain how rugged it was. In both areas, however, the sharp features seemed muted by some overlying material; perhaps it was windblown sand. Utopia had a slight advantage in that the covering appeared to spread more evenly over the area of a possible landing ellipse.

5.11   The search now moved to a region far around the planet, the region of Utopia. Mosaics were obtained of this area. At the top right-hand corner of this mosaic is the crater Mie, which is about the same size as the lunar crater Copernicus. The southern part has peculiar features reminiscent of the Yukon seen from a high-altitude jet in winter. But closer to the crater Mie, the area appears to be blanketed with fine material that has smoothed out much of the roughness.

Now it came to a vote by all those present at the meeting: Should a landing be attempted, or should further searches be made? The decision was to go ahead and attempt a landing. Next the site certification was voted on. The vote was two to one in favor of landing in Arcadia. But Jim Martin had to make the final decision. He and his executives met for four hours in a closed meeting separate from the other, more general meeting. Martin decided on Utopia, and the announcement was made. He felt that it was the safest place for a landing. The Utopia site at 47.9 degrees north, 225.9 degrees west was almost halfway around Mars from the landing site of the first Viking.

Where to Land—The Blandlands

The "go/no-go" decision meeting for the second landing was held on September 3. Jim Martin said that separation would take place at 12:40 P.M. PDT on schedule. But one of the radar beams on the lander developed a problem, and Martin decided to land with three beams instead of four as used on lander 1. Martin explained that he did not want to risk a faulty beam feeding erroneous information into the navigation computer of the lander. Landing was still expected to take place at 3:58 P.M. PDT that day.

The lander separated from the orbiter on schedule, but immediately there was trouble. The data stream from the orbiter to earth ended abruptly. A power failure had caused loss of control on the orbiter. A backup system was brought into operation automati-cally, but the orbiter had lost its precise orientation in space; now the high-gain antenna pointed away from earth, thereby stopping communications. The orbiter then automatically stabilized itself and ceased operations while it waited for instructions from earth.

This shutdown of the orbiter was a precautionary measure built into its system to prevent it from doing anything that could damage it further. It was unfor-

5.12 There was much scientific debate about whether a landing should be made at Utopia or in Arcadia (west of Alba Patera). Here Jim Martin, flanked by Michael Carr on the left and Tom Young (Mission Director) on the right, takes a vote during the final decision process. To the right of Young are Gentry Lee and Gerry Soffen.

tunate that this safety feature had to be activated at such a crucial time of the mission. Commands were sent quickly to switch the orbiter's communications from the high data-rate mode to a low data-rate mode. In this way mission control and the orbiter could talk to each other without having the high-gain antenna pointed directly toward the earth.

After agonizing minutes some engineering data began to be received again at JPL. Time was becoming critical. The lander was on its way to the surface of Mars, and although engineers now understood the problem with the orbiter, there was no time to issue commands to change communications back to the high data-rate mode before the landing. The only information that could be received at JPL was certain low-rate engineering signals about the status of the orbiter.

So the second Viking lander went down to the Martian surface without sending any data back to earth. The orbiter was still receiving data from the lander, but it was storing them in its memory. It could not send them to earth until the high-gain antenna was again directed properly.

The computers of the orbiter had been commanded to turn on and receive high-speed data from the lander as soon as it reached the surface. This command had been sent prior to separation but the shutdown canceled it, and the activity had to be recommanded. The orbiter's relay communications system also had to be commanded to record and relay the stream of data from the lander after a successful landing. This change to a high rate of 16,000 bits per second would appear in the engineering data flowing from the orbiter to earth; this would, indeed, be the first sign on earth that the lander had touched down successfully.

At the Jet Propulsion Laboratory the many people who had gathered for this second landing quietly watched the minutes flick by on the digital clocks. Everyone was anxious when the expected time of touchdown passed without any indication from the orbiter that the data rate had changed. Finally, almost a minute late, a successful touchdown was announced over the communications net. Cheers went up through the control center and elsewhere in the JPL complex. Viking 2 had made it safely to Mars. But there was no immediate picture playback this time, as there had been with the first landing. Orbiter 2 had to be put back into normal operation before the pictures being taken of the Utopian landscape could be seen on earth, and before the data obtained during entry could be relayed to earth from the memory aboard the orbiter.

6.1 The first picture from the surface of Mars (consisting of three frames) showed a footpad resting securely on the soil of the Red Planet. The vertical streaks at the left side of the picture were thought to be caused by dust raised at the time of the landing.

Chapter 6

Touchdown to Hibernation

July 20 is a historic date in space exploration—in 1976 Viking 1 landed, and seven years earlier to the day, man first stood on the moon. In the future, some of us thought, July 20 might be more important than any other celebration; it might be celebrated annually to commemorate mankind's ability to set foot on other worlds and to reach other planets.

When the cheers and back-slapping and toasting subsided at the Jet Propulsion Laboratory, everyone began to gather around the TV monitor screens again. One of Viking's cameras had been automatically scanning the surface of Mars close to a footpad of the spacecraft. The mirror within the metal cylinder had nodded up and down, gradually moving across the scene to record a picture of Mars as a series of vertical strips. The differences in light and shade in each strip had been passed as a string of numbers back to earth, where they were retranslated by the computer into a picture.

About half an hour after the landing, this picture was to appear on the TV monitors placed at many locations throughout the Laboratory. Everyone watched intently. Would there be plants? Would there be rocks? While we waited I thought about the fantastic achievement of the landing on Mars. It had been only 17 seconds later than predicted, after a journey of

400 million miles (650 million km) in the 335 days since the big Titan/Centaur had lifted the spacecraft from Launch Complex 41 at the Kennedy Space Center in Florida. Viking 1 had touched down only 20 miles (33 km) from its aim point at 22.46 degrees north latitude and 48.01 degrees west longitude.

The first picture sequence started as expected; slowly line by line the first frame built up on the TV screens. We were looking at the surface of Mars as clearly as if we had been sitting on the lander. Scientists and laymen alike were elated at the clarity. We could see rocks of many kinds, and to the extreme left a vertical dark streak looked like a shadow. Dr. Thomas Mutch, leader of the imaging team for the Viking lander, commented that it was probably a shadow cast by a cloud of dust raised by the landing.

A short while later a second picture replaced the first on the screen, and then was replaced by a third. On the third frame we saw the footpad of the Viking 1 lander resting securely on the Martian soil (figure 6.1). Some soil fragments had been disturbed and lay on top of the footpad. But it was clear that the footpad had not penetrated deeply and that the soil of Chryse Planitia could support a spacecraft.

The lander's cameras were preprogrammed to take two picture sequences on this first day. Viking 1

touched down at 5:12 A.M. PDT, but the local time on Mars at touchdown was 4:13 P.M. Sunset on Chryse would occur only a few hours later. Therefore the first picture sequence described above was planned to show the lander's footpad and the area around it. This provided scientific information about the details of the Martian soil. A second sequence was a panoramic view in front of the lander to show the general landscape at the landing site. Then if Viking did not survive its first night on Mars, the mission would still have given the scientists good views of the Martian surface. Earlier Soviet spacecraft had been unable to do this because they had failed almost immediately after landing.

The panorama arrived as a series of frames which, placed side by side, covered a full 300 degrees of the Martian horizon (figure 6.2). As the panorama unfolded on the screens, we saw the meteorology boom of the spacecraft and an area of dunes. In the far distance, on the horizon, we could see light-colored rock masses illuminated by the setting sun, which was shining directly on their high cliffs. As the picture extended, a fantastic Martian landscape materialized, an undulating rock-strewn countryside completely unlike any lunar landscape seen from the Apollo landings. This scene was more like the western deserts of the United States. And the sky was pale, not black like that seen from the moon.

The rolling nature of the countryside held everyone spellbound. Were those dark markings vegetation? There's a rock that looks like a cylinder! People imagined all sorts of things among the jumbled rocks. But when copies of the pictures became available soon afterward, it was quite apparent that no vegetation existed on the surface of the Chryse basin. It was a barren, rocky world. Surprisingly, there were no signs of craters, none of the small craters that dot the lunar plains and are so prominent in all the pictures of the lunar surface taken by Apollo astronauts.

The surface in Chryse consisted of rocks and finely granulated material; whether it was sand or dust could not yet be determined. In the foreground there were small rocks with flattened faces and angular facets. Several larger rocks showed irregular surfaces with pits, and at least one of the rocks had intersecting linear cracks. Associated with several of the rocks there were fillets of fine material between the rock and the surface, and streaks on the surface—evidences of material having been transported by the wind.

At a press conference soon after the landing, members of the Viking team and NASA officials commented on the landing and the first results of this expedition to the Red Planet. Jim Martin proudly said, "This is the happiest day of my life—I've lived a long

Touchdown to Hibernation

6.2 The first panorama of the Martian surface in Chryse Planitia, taken on the afternoon of Viking's first day on Mars. The view is toward the south-southwest but extends from the east-northeast on the left to north on the right, covering 300 degrees. The rolling nature of the rock-strewn terrain is apparent.

time for this, and I want to thank all the people who made it possible. There must be 10,000 people in this country that deserve a part of the credit." Martin continued, "I don't plan on luck. I believe that most of it you make; and just to give you an example, it's people doing that extra job. We had a little concern over the radar in the preseparation checkout, and yesterday some 50 guys in Denver turned to and ran probably a dozen tests on those radar altimeters inside of 8 hours. And that convinced us we didn't have any problem whatsoever. It's that kind of dedicated, undirected [spontaneous] work and effort that, as far as I'm concerned, is what you use to replace luck."

Thomas Young, the mission director, who had coordinated the teams of operations people responsible for the safe landing, was very pleased: "All indications are that everything [on the lander] performed extremely well. I think truly today the search begins, and I am proud to be a part of it. The quality of the first photographs exceeded my expectations. Mars today demonstrated that it is photogenic."

Dr. Robert Kraemer, director of planetary programs at NASA headquarters, admitted that "the moment it touched down was a heart-stopper."

Contact with the lander ended about 15 minutes after touchdown as the motion of the orbiter carried it out of range of the landing site, and the rotation of Mars carried the lander to the far side of Mars as viewed from earth. During this silent period, scientists were striving to interpret the mass of scientific data that had been returned to earth during entry. Dr. Alfred Nier, leader of the entry science team, and other members of the team discussed their findings later on that first day. The most important findings were that the atmosphere contained 3 percent nitrogen, whose presence could not be detected from earth, and that the amount of argon—the gas that could affect the operation of the gas chromatograph–mass spectrometer on the surface—was less than 2 percent (later the percentage was revised upward to 2.5 percent). The argon would not pose any problems.

In the Martian upper atmosphere, the molecular oxygen ion with a single electrical charge (O_2^+) was found to be the major constituent at an altitude of 81 miles (130 km). The carbon dioxide ion (CO_2^+) was one-ninth as abundant. Direct measurement of this electrically charged molecular oxygen was important to our understanding of the photochemical process

Touchdown to Hibernation

occurring in the upper atmosphere of Mars. It supported a theory that the carbon dioxide ions produced by solar ultraviolet radiation undergo further reactions in Mars' ionosphere to produce more stable oxygen molecules and carbon monoxide.

On both Mars and the earth, the gases of the atmosphere were released from the interior of the planet through volcanoes and other means of venting. On earth, much of the vented water vapor condensed and became the oceans, and most of the carbon dioxide is now part of limestone rocks, coal, and petroleum, and the tissues of living things. On Mars too, much of the vented gases may have gone into nonatmospheric reservoirs.

The data obtained during entry provided information about the various isotopes of the Martian atmospheric gases. This is highly important in finding out what Mars' atmosphere may have been like in the past. Proportions of isotopes of both carbon and oxygen are similar to those of the terrestrial atmosphere. The atmosphere of Mars is enriched, however, with the heavier, rare isotope nitrogen 15 relative to nitrogen 14. The whole subject of the significance of enrichment is extremely complex.

The ratio of nitrogen to argon 40 (produced by radioactive decay in the crustal rocks of planets, mainly from potassium 40) was calculated for Mars and compared to the ratio on earth, allowing for the difference in masses of the two planets. From these calculations the experimenters concluded that, over time, at least 7 millibars worth of nitrogen had been released on Mars. Today the Martian atmosphere contains only about 0.2 millibars of nitrogen; some has escaped into outer space, and some may have been deposited in the loose surface rock (the regolith). This would be important for life processes. This result, said Michael McElroy, an entry-science team member, might have been the equivalent of a shower of "fertilizer" falling from the atmosphere of Mars into the soil.

On earth there is about 1 percent argon in the atmosphere, compared with 1.5 percent in the Martian atmosphere. To account for this additional argon, there must have been much larger quantities of nitrogen, water, and carbon dioxide on Mars in the past, and most of these must still exist on the planet; the question is, Where? This question would be hotly debated in the following months.

Temperatures of the upper atmosphere derived from Viking experiments were considerably lower than those inferred from measurements of the airglow (arising from release of energy in the upper atmosphere), made by earlier Mariner spacecraft. This difference might occur because the Viking landers penetrated the atmosphere when the planet was near its maximum distance from the sun, while earlier measurements were made when Mars was closer to the sun.

Temperature profiles (figure 6.3) traced by both spacecraft showed wavelike patterns above 19 miles (30 km), an unexpectedly complex thermal structure. Also, results for the second Viking differed from those obtained with the first. The temperature waves were thought to record several days of the thermal cycle at the planet's surface, moving upward at about 0.6 miles per hour (1 km/hr) by radiative heat transfer, i.e., the movement of heat upward into the atmosphere by radiation as contrasted with actual movement of atmospheric gases carrying heat with them (convective transfer).

Throughout the atmosphere, the temperature was well above that at which carbon dioxide condenses. Clouds and mists on Mars must consist mainly of water, not of carbon dioxide. Atmospheric pressures were measured by both Vikings during entry and agreed with those measured after touchdown.

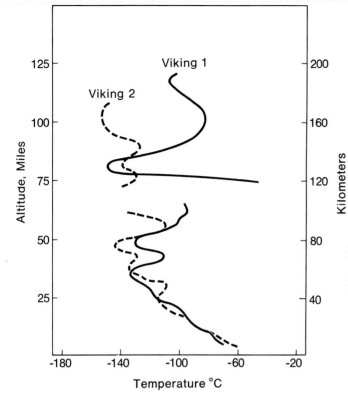

6.3 The temperature of the Martian atmosphere was obtained during entry of the two Viking spacecraft and is shown here plotted against height above the surface. Waves in the temperature profile are thought to be caused by daily variations moving upward.

6.4 The meteorology boom extended its sensors, which report the daily weather on Chryse. The sensors are placed about 4 feet above the surface and measure atmospheric pressure, temperature, wind velocity, and wind direction.

Changes in the density of the Martian atmosphere with height were also measured. At high altitude it was more dense than expected, but at the surface it corresponded with predictions; namely, 0.00112 pounds per cubic foot (0.018 kg/m³) compared with earth's atmospheric density of 0.08 pounds per cubic foot (1.29 kg/m³).

In the days after the landing, the scientific instruments on the lander began to send data from the surface. Nearly everything worked perfectly, but the seismometer, which had been protectively restrained in flight and during the landing by a caging mechanism, refused to become released. Despite repeated commands, it stuck tight. Weeks of engineering analysis showed that probably nothing was wrong with the instrument itself, but rather a grounding wire that carried the current to release the caging mechanism was not connected.

The meteorology boom had unfurled, as seen on the first panorama (figure 6.4), and the sensitive instruments mounted at its tip had sent the first weather reports from the Red Planet. These were passed on in a bulletin from Dr. Seymour Hess, the leader of the meteorology team, on July 21: "This is the first weather report from Mars in the history of mankind, the weather in Chryse on sols [days] zero and one. Light winds from the east in the late afternoon, changing to light winds from the southwest after midnight. Maximum wind was 15 miles per hour. Temperature ranged from −122 degrees Fahrenheit, just after dawn, to −22 degrees Fahrenheit, but we believe that was not the maximum. Pressure is steady at 7.70 millibars."

The X-ray fluorescence spectrometer for the inorganic chemistry experiment was calibrated soon after the landing, although it would not receive a soil sample for analysis until later. In this process it provided more information about the Martian atmosphere and confirmed the percentage of argon as measured by the entry-science experiments.

The first color picture of the Martian surface was shown at a press briefing by Dr. Carl Sagan, a member of the lander imaging team. The Chryse "plain of gold" was not gold, but pink. The general color impression was of reddish soil and rocks, with some greenish areas and deposits. The picture showed a blue sky, but Dr. Sagan explained that this arose from a faulty color mixing in making the picture from the data. Martian sky, he said, was really pink. And when the audience laughed at this announcement of a pink sky, he chided them for a typical earth chauvinistic response.

Dr. James Pollack of the Ames Research Center, in discussing the color of the sky and its meteorological implications at the briefing, said that "the sky is extremely bright, as bright or brighter than the surface itself. We immediately know that most of the brightness we see is due to scattering of particles [in the atmosphere]."

In answer to questions on why the sky of Mars should be blue, or gray, or pink, Dr. Pollack replied, "I think . . . the color of the sky . . . will basically depend upon the competition between how many soil particles are there and how many water-ice particles. At times, in the places where water-ice particles form, then the sky will tend to be more grayish in color. However, I think we're seeing something that's more than just a localized dust event, but something that really is a global dust cover. I would expect that as long as there weren't water-ice particles, at most places on Mars the sky would be pinkish in color."

In subsequent days, new color pictures were produced on which the color was matched to known parts of the spacecraft visible in them—a color chart, orange cables, and the American flag. The Martian scene was without doubt a pinkish, red world.

The rocks were also pink. The lander imaging team discussed the meaning of Mars' color at this same press conference. Said Dr. Thomas Mutch, leader of the team, "The contrast between Mars' colors and the moon's colors shows that Mars and the moon have entirely different chemical histories." Dr. Alan Binder, a team member commented:

> The color we are getting matches that which the astronomical observations show, so the color is quite true. The rocks are coated in these pictures with a reddish material, the fines [fine particles] are coated with the same material. We obviously do not know whether this holds for the rest of Mars; we've only looked at one place in terms of the high resolution that we have with the surface cameras, but I'm fairly confident that the model (a limonite surface stain) will apply to the rest of Mars.
>
> Now limonite itself is, in everyday terms, simply rust. There are two minerals which . . . are the end members of a suite of materials. Hematite is fairly red, quite red in fact; and goethite is a yellowish mineral. Hematite is pure iron oxide; goethite is a hydrated iron oxide. Limonite is an amorphous mixture of these two minerals in varying proportions. If you have a limonite which is rich in hematite, it is obviously fairly red; if the limonite is fairly rich in goethite, it is orangish or more yellowish. You can find quite a range of these materials on the earth in terms of the hues, so we can, by picking the proper source region, probably match the surface of Mars without any difficulty.
>
> I would make one remark, though; the lithology [gross physical character] of these rocks in the area [where] we collected these samples is a granite or granite-gneiss. This is a very different parent rock than what we expected to find on Mars. Granites are a light-colored rock.
>
> We have really quite a variety of different parental rocks in this region [of Mars], all of which are stained. This is very exciting from the standpoint of our attempts to determine the lithology of the surface. It will tell us quite a bit about the geochemical development of the crust of Mars.
>
> On earth, in an arid environment you develop this stain by oxidizing and hydrating the mafic minerals in the rocks. Mafic minerals are those minerals which are dark; they are iron rich. What effectively happens is the minerals are chemically broken down to produce rust. This surface is produced on the earth only in a fairly arid envi-

Touchdown to Hibernation

ronment. We can only match the colormetric properties of Mars with those of materials which were formed in very arid environments.

Dr. Binder's comments seemed to suggest that there had been little free water on Mars during its recent history.

Meanwhile, although the photography was going well, a serious problem had developed with the surface sampler. This scoop, on the end of a long extendable arm, was needed to pick up samples of Martian soil for the important biological experiments and for chemical analysis of the surface materials. On Thursday July 22, at 1:00 A.M. PDT (about 10:30 in the morning in the Chryse basin on Mars), the computer of lander 1 issued commands to test the soil sampler mechanism in preparation for actual sampling that was scheduled to take place in another few days. The sampler and its arm were told to go through a sequence of 17 movements.

Telemetry data sent back from the lander indicated that only 13 of the movements took place, and then the sampler arm locked up. Project officials thought the problem might be in the electrical system rather than the mechanics of the sampler arm. A team of engineers operated the sampler arm on the test lander spacecraft at the Jet Propulsion Laboratory (figure 6.5), trying out various theories of what might have happened. That evening, Jim Martin reported that after a very busy day he believed the cause of the malfunction and a solution to it had been found. "It turns out, contrary to my expectation, not to be an electrical problem. The most probable cause is a mechanical problem. It is related to the fact that there is a locking pin that is part of the shroud latching system. It was supposed to drop to the Martian surface during the boom extension which was part of the shroud ejection sequence." The metal cannister called a shroud covered the vital collector head and its scoop to protect it during the voyage to Mars and during the landing. It was held on by a latch which,

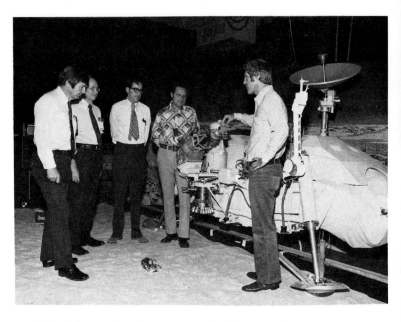

6.5 A team of engineers operated the sampler arm on a lander in the atrium at the Jet Propulsion Laboratory, testing out various theories of what had caused the sampler arm to stop operating on Mars.

when released, allowed the shroud to be ejected from the collector head. "It now appears that the extension that had been commanded into the sequence was not long enough to allow this pin to fall free."

By the next day, July 23, engineers using the science test spacecraft (a replica of the lander) at JPL and the proof test lander (another replica) at Martin Marietta had successfully tested a plan to free the soil sampler arm. Commented mission director Tom Young, "We believe we understand the cause of the problem. We believe we've identified the most probable fix, . . . to extend the sampler boom . . . about 13 inches and allow a pin . . . that was part of the lockdown mechanism, to fall free. We believe that it is that pin that has been jammed in the surface sampler."

A new command sequence was sent to extend the boom. Tom Young expressed his confidence that exploration could begin. "We are now in a condition that all of the deployments that are necessary to move into the next phase of the Viking exploration of the surface of Mars have been accomplished. The

6.6 High-resolution pictures revealed fascinating details of the Martian surface. This view looks northwest from Viking 1. It shows large boulders, including the dark rock that was later called "Big Joe." This rock is about 10 feet wide and 3 feet high. It is capped by lighter material.

nas on the biology instrument and on the gas chromatograph—mass spectrometer instrument [organic analysis] have been opened. They're ready to receive samples." But it would be several days before the sampler arm went through the motions commanded to it.

During this critical period there were also some minor problems. One of two (redundant) radio receivers in the lander, through which commands were received from earth, did not work properly. Also, the lander's radio transmitter operated at a lower power mode than that commanded. Although troublesome, these problems did not cause any loss of data.

Meanwhile, more pictures of the Martian surface were coming back to earth (figures 6.6–6.8). Com-

menting on them, Dr. Elliott Levinthal, a member of the lander imaging team, described a process used to look for movement on the Martian surface. Normally the lander camera moved its view sequentially in a horizontal direction so that the scanning lines were side by side, thus making up the complete picture as a series of vertical strips. But the camera could be commanded to repeat one line without moving. When such a picture was reconstructed on earth, it showed the same line repeated side by side over the width of the picture (figure 6.9). The picture looked

Touchdown to Hibernation

6.7 This picture was taken by Viking 1 during its third day on Mars. It shows numerous angular blocks ranging in size from a few inches to several feet. The surface between the blocks consists of fine-grained material; some accumulates downwind of obstacles. The large block on the horizon is about 13 feet wide. Some of the flat areas may be bedrock protruding through the debris.

6.8 One of the most spectacular pictures returned by Viking 1 was this panorama, showing dune fields beyond the meteorology boom. The dramatic early morning lighting (7:30 A.M. local time on Mars) reveals subtle details and shading. The sharp dune crests indicate that the most recent wind storms capable of moving sand over the dunes blew from the general direction of upper left to lower right. Small deposits downwind of rocks also indicate this north-northeast wind direction.

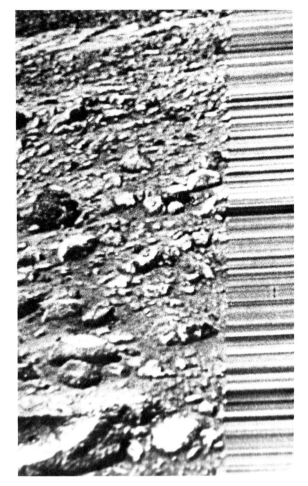

6.9 Several pictures were made in a constant-scan mode whereby the mirror of the camera continued to nod up and down but did not move horizontally. The picture smears into a series of horizontal lines. Any changes to these horizontal lines would represent something changing on the surface. No such changes were observed.

like a series of horizontal bands. If anything changed during the taking of the picture, the movement or change would show up as a difference across one or more of the horizontal bands. At the Viking lander 1 site the experiment showed no change—no dust blew by, no rocks moved.

The mass spectrometer of the organic analysis experiment, freed of any chance of damage from too much argon in the Martian atmosphere had been commanded to sample the air at the landing site, looking for rare gases. Its initial results confirmed that the atmosphere of Mars near the surface consisted of 94 percent carbon dioxide, 2–3 percent nitrogen, 1–2 percent argon, and about 0.3 percent oxygen.

This was the first time that the nitrogen in the Martian atmosphere was measured directly, and later the spectrometer detected the rare gases krypton and xenon, krypton being more abundant than xenon. The ratio of xenon isotopes 129 and 132 was found to be enhanced on Mars compared with the earth. The excess of krypton over xenon in the Martian atmosphere differed from that in a class of crumbly meteorites, called carbonaceous chondrites, which are believed to represent very early material of the solar system. It more closely matched that of the earth's atmosphere. The xenon deficiency on earth is attributed to absorption of the gas in shale, a sedimentary rock. Similar processes might have taken place on Mars if there were earlier periods of plentiful water.

Dr. Alan Binder, of the imaging team, reported on the first stereoscopic (three-dimensional) pictures, made possible when the second camera was operated. Viking now had two eyes and could see with depth. This second camera was protected until after the first pictures had come back, in case clouds of dust were blowing about on the surface when the spacecraft landed. Such dust would abrade the window of a camera. As revealed in the stereo pictures, the topography at the site was quite rolling, said Dr. Binder. There were dunes of both light and dark material, and groups of fairly large rocks. In some places, bedrock penetrated the rocks and fine material. Many rock types were visible in the pictures—coarse-grained, smooth, weathered, even hollowed-out rocks similar to terrestrial rocks with gas blisters that are found on the top of lava flows. Some lag gravel could be seen. Also called desert pavement, it results when fine material is blown away by the wind, leaving coarse particles beneath, and it is common in terrestrial deserts.

When I walked into the press room at JPL early on the morning of July 24, colleagues pointed to a pic-

6.10 The letter B, or perhaps a figure 8, appears to have been etched into the Mars rock at the left edge of this picture, taken July 24. It is believed to be an illusion caused by weathering processes and the angle of the sunlight on the rock.

ture from Mars displayed on the TV monitors (figure 6.10). "Have you seen that?" they asked. "What do you make of it?"

Several correspondents, clustered around the monitor screen, were looking closely at a fairly large rock near the left side of the picture. Dr. Binder had spotted it when the picture was received earlier that morning. The rock was intriguing because it looked as if a letter B was carved on it, and other letters too. "Rocks can take on very interesting shapes; you can find every imaginable thing you want," commented Dr. Binder. He added that the letters were effects of shadows on the rough face of the rock. Visions of a Martian Rosetta stone faded rapidly. The lettered rocks of Mars joined the Schiaparelli/Lowell canals as figments of human imagination.

By July 25, JPL received pictures confirming the commanded movements of the sampler boom. The pin had dropped to the Martian soil. All was now

ready for the soil sampling on sol 8, the eighth day on Mars.

During this period, preparations were made to certify that the soil sampling site was suitable. Before the lander separated from the orbiter to go down to the surface, its computer had been instructed for a full 90-day mission, including sampling soil from a site in front of the lander. If for some reason the lander could not have been commanded again from the earth, the soil sampler would have automatically attempted to pick up a sample.

With a successful lander, the science teams had an opportunity to choose the best site available. Examination of the pictures showed that the preselected digging spot was quite unsuitable. Stereo pictures of the surface were used with a computer program to trace contours of several potential sites.

A new site was finally selected at the base of the drifts of fine-grained material seen at the left of the first panorama picture. Commands were sent to the lander to start digging for a sample at 1:00 A.M. PDT on July 28. The soil sampler was told to deliver the first load to the biology experiment, the second load

Touchdown to Hibernation

to the inorganic analysis experiment, and the third load to the organic analysis experiment. The lander's cameras were also commanded to photograph the sampling process.

The first picture returned that day showed that the surface sampler had successfully scooped up soil from Mars (figure 6.11). By noon, telemetry data confirmed that one of the three tests in the biology instrument had started, indicating that it had successfully received soil. At 2:00 P.M., Viking mission control reported that "the soil has been successfully loaded in two of the three instruments on board the spacecraft. However, there are indications that one of the three instruments did not get its sample." The indicator on the organic analysis experiment had not come on. There were four possibilities: the surface sampler had not delivered an adequate sample, the grinder motor had not operated to grind the soil into the right size, the material was not fine enough or was too sticky to pass through the seive, or a sensor in the instrument that detected a suitable level of soil within the sample chamber had failed to operate correctly. So the important organic analysis experiment was held up until a new sampling could be commanded.

In the meantime, the first significant data about the Martian soil were obtained with the inorganic (X-ray fluorescence) chemical analysis experiment. The major constituents of the soil were iron, calcium, silicon, titanium, and aluminum. Although iron oxides were present, the sample was not pure limonite but seemed rather to have a limonite veneer. A surprise was that the trace elements rubidium, strontium, and zirconium were present in extremely low concentrations compared with the earth. Also, the experiment showed that the ratio of calcium to potassium in Martian soil was higher than in terrestrial soils.

Project management decided to try to pick up another sample for the organic experiment on August 4.

6.11 The trench shown on this picture, returned by Viking 1 on July 28, was dug early on sol 8. It is about 3 inches wide, 6 inches long, and 2 inches deep. Doming of the surface at the far end shows that the granular material is dense. It behaves somewhat as moist sand does on earth.

To their distress, the sampler boom stuck again. By this time the biology experiments were producing startling results, as described in the next chapter, and a major question was whether these results indicated biology on Mars or an active surface chemistry that was not biological. Everyone urgently wanted to know if there were organic compounds—remains of Martian microbes—in the soil of Mars. Obtaining an answer was the task of the organic analysis experiment. Project management decided that a test should be run in the hope that some material had been deliv-

Touchdown to Hibernation

6.12 Lander 2's first picture of the Martian surface at Utopia was taken within a few minutes of touchdown on September 3, 1976. The shading at the left is again thought to be dust kicked up by the landing. There are boulders of 4 to 8 inches, some spongy like lava and others fluted by the wind. Many of the pebbles have tubular or platelike shapes, suggesting that they might have been derived from layered strata. The grooved boulder just above the footpad has a scraped surface, which suggests that it may have been struck by the footpad and overturned. Fine-grained material was kicked up by the descent engines and deposited in the concave interior of the footpad, as with lander 1.

6.13 Panorama at the Utopia landing site. This presents quite a different scene from that in Chryse. The rocks are smaller and the terrain is very flat. There are no dune fields like those at Chryse, even though observations from orbit had suggested that Utopia was covered with dunes.

ered into the test chamber on the earlier attempt. An organic analysis cycle was started on August 6, heating the chamber to 392 degrees Fahrenheit (200°C). The next day, data from the experiment were telemetered to earth. They confirmed that there was a sample of soil in the test chamber, but the results were unexpected: the sample of Martian soil contained no detectable organic molecules. The scientists anxiously awaited another confirmatory test.

Subsequently the problem with the sampler boom was overcome by operating it during a warmer part of the Martian day; operations resumed on the surface on August 11. In the following weeks, samples were

obtained from other sites, rocks were nudged and pushed, and a full program of investigating the surface of Mars was completed by Viking 1 before it was placed in hibernation in mid-November for the period of solar conjunction.

Viking 2 had made its successful landing on September 3, in the Utopia Planitia at 47.96 degrees north latitude and 225.77 degrees west longitude. The first pictures were delayed for over 10 hours by problems with the orbiter and its high-gain antenna, as described earlier, but when they arrived they were as awesome and exciting as those of Viking lander 1.

The first picture showing the footpad (figure 6.12) was similar to that from Viking 1, although the rocks appeared more pitted, but the panorama picture revealed a very different landscape from that at Chryse. The panorama swept around 324 degrees of the view from the lander, almost a full circle (figure 6.13). It

Touchdown to Hibernation

revealed a surface strewn with rocks to the far horizon, which was much flatter than at Chryse. These rocks ranged in size up to several meters but did not show the variety of sizes seen at Chryse. Many of the rocks appeared to be smooth. Others had pitted surfaces and resembled fragments of terrestrial volcanic lava. Some of the rocks seemed to be grooved, possibly by windblown sand or dust. In contrast to the Chryse site, there were no sand dunes despite the expectation based on orbital pictures that the site was in a dune field. There were, however, deposits of fine-grained material between the rocks.

An unusual feature seen in the panorama was a channel running from the left across the mid-distance to the foreground at the middle of the picture. This channel was relatively free of rocks. The speculation was that it might be a dry stream bed, or a partially filled crack in underlying bedrock.

On September 6, a photo of disturbed rocks near the footpad of lander 2 showed that the protective covering for the lander's soil-sampler head had been successfully ejected. Later, the shroud was photographed on the surface. Another piece of good news from this lander was that the seismometer had uncaged when commanded. Viking 2 had an active seismometer on Mars.

The first weather report from Utopia was similar to that at Chryse: winds were light, between 3 and 11 miles per hour (5 to 18 km/hr), and temperatures somewhat cooler. The first noon temperature recorded was −31 degrees Fahrenheit (−35°C), and the nighttime low was −128 degrees Fahrenheit (−89°C). Pressure was 7.78 millibars.

Analysis of the atmosphere at the Utopia site gave the same composition as at Chryse, as expected.

As with Viking 1, the delivery of Martian soil started on the eighth day after landing. Again, problems arose with the scoop. It stopped rotating during the soil delivery sequence just before it could deposit a load of soil into the hopper of the inorganic chemistry instrument. Fortunately the biology instrument received its sample and proceeded with its first cycle of analysis.

At JPL, engineers tried to figure out what had gone wrong with the scoop this time. One theory was that a pebble had stuck in the hinge of the backhoe. But a picture of the sampler returned on September 15 showed that this had not happened. Engineers then said that a faulty position switch might have caused the sample head to stop operation. Commands were sent to override the switch. Three days later the scooop operated successfully and dumped a load of soil into the inorganic experiment. Sampling of surface material to look for organic compounds was delayed, however, to allow scientists to find a better area for taking the sample. The site from which the first biology sample of lander 2 had been dug was not considered a good one for organic molecules, since the results from the biology experiments were almost identical with those from Chryse, and the Chryse sample had not contained detectable organic material.

On September 25, the surface sampler obtained soil for the organic analysis from the "Bonneville Salt Flats," a small area surrounded by rocks. The digging of the sampling trench required great precision to avoid bumping into rocks, and progress was closely watched by the camera (figure 6.14), but the pictures could only be seen after the action had taken place.

As scientific investigations continued with lander 2, flight controllers sent commands on October 2 to dig more soil samples. Late on that day the soil collector gouged a new trench to obtain material for the second cycle of biology investigations. On October 3, the sampler dug for small rocks and gravel needed for the inorganic experiment. On the following day, the

Touchdown to Hibernation

6.14 Operation of the surface sampler was closely monitored when it dug a trench in an 8 by 9 inch area surrounded by rocks. The exposed thin crust was a prime target for organic materials. *At left,* the sampler scoop touched the surface, missing the rock at upper left by a comfortable 6 inches. The backhoe has penetrated the surface about half an inch. The scoop was then pulled back to sample the desired small area and (*second picture*) the backhoe furrowed the surface, pulling a piece of thin crust toward the spacecraft. The initial touchdown and retraction sequence was used to avoid a collision between a rock in the shadow of the arm and a plate joining the arm and the scoop. The rock was cleared by about 3 inches. The third picture was taken 8 minutes after the scoop touched the surface. It shows that the collector head has acquired a quantity of soil. With the surface sampler withdrawn (*right picture*), the foot-long trench is visible between the rocks. This trench is 3 inches wide and about 2 inches deep. The scoop reached to within 3 inches of the rock at the far end of the trench.

6.15 On October 8, 1976, a rock-pushing attempt was successful. The irregular-shaped rock was pushed several inches by the lander's collector arm. This rock, named "Badger," was moved to the left of its original position and left cocked slightly upward. A soil sample from the area initially underneath the rock was taken a short while later and delivered to the experiment that was searching for organics on Mars. Scientists hoped that the sample might have been protected from solar ultraviolet radiation so that life forms or organics would be present.

sampler attempted to nudge a rock to obtain a sample from beneath it for organic analysis. The organic analyses from both landers had so far produced no evidence that organic molecules or microbes existed on Mars, and the experimenters thought that a sample from beneath a rock, protected from solar ultraviolet radiation, might contain organic molecules. The selected rock refused to budge.

On October 8, a second rock was pushed by the sampler arm (figure 6.15). This time the effort succeeded. The sampler arm then reached out and dug a trench in the soil exposed by moving the rock and delivered this sample to the organic experiment for analysis.

By October 18, the organic experiment tests were completed on the soil sample from under the rock. There were still no organics. Everyone was puzzled by these results; all tests for organics on Mars had yielded negative results. Moreover, the sample from under the rock showed no extra evolution of oxygen, water, or carbon dioxide beyond that expected from samples taken from exposed parts of the Martian surface. Speculations that rocks might protect the surface from whatever atmospheric and radiation effects might be responsible for the soil's active chemistry received a severe blow. There was still a possibility, however, that solar ultraviolet light had destroyed organic molecules between the moving of the rock and the collection of the sample.

Encouraged by the success of the rock-pushing activity, project manager Jim Martin now planned "to push two rocks on October 19, to find one that moves and to perhaps have a choice as to which rock we want to dig under for biology. We will move it out of the way then, and dig under it in a very short sequence to minimize the exposure of the soil to the exterior world . . . on the 25th of October."

By October 21, scientists had picked the next rock to move. They named it the "notch" rock. Five days later the sampler arm was extended, the rock was pushed aside, and soil was collected from under it and transferred to the input hoppers of the three life-detection experiments. All this sampling took place in a few hours to prevent ultraviolet damage to possible life forms in the soil sheltered by the rock. But the results of these analyses were very similar to those from the exposed surface material.

A serious problem remained in the inorganic analysis. The scientists wanted to analyze pebbles to get a better idea of the elemental composition of the local rocks, but no pebbles had been collected. Although attempts had been made to collect them, the science team concluded that the samples were actually cemented fine materials rather than pebbles. The samples of "fines" were more likely to give the average composition of the Martian surface, because they had undoubtedly been blown and mixed by winds during the great Martian windstorms.

As the days passed, the weather conditions hardly changed at either site. At Chryse, temperature highs each day were about -19 degrees Fahrenheit ($-28°C$), and lows about -116 degrees Fahrenheit ($-82°C$). At Utopia, temperature highs were -23 degrees Fahrenheit ($-31°C$) and lows about the same as at Chryse. However, the wind patterns produced some interesting new information about Martian meteorology. Daily and twice-daily fluctuations of the Martian atmosphere were recorded and believed to be evidence of atmospheric tides, caused by solar heating cycles. The daily change in pressure at Chryse amounted to about 0.1 millibar, whereas at Utopia it was only 0.02 millibar. At both sites the daily pressure variation gave greatest pressure each day at about local midnight, and the lowest pressure at about 4:30 P.M. While these pressure variations may seem small, they actually represent a large proportion of the atmospheric pressure on Mars, and proportionately they are about five times as great as daily pressure variations in a desert environment of earth. These pressure variations indicate that the atmosphere of Mars is subjected to atmospheric tides and wind systems of much greater intensity than those of the earth.

As time progressed, the mean atmospheric pressure fell gradually, indicating that some of the carbon dioxide was being removed by condensation at the growing north polar cap. As expected, pressures reached a minimum at both sites about the third week in October 1976, close to mid-winter in the northern hemisphere of Mars, then began to rise slowly. The minimum at Chryse was 6.81 millibars (compared with the pressure at time of landing of 7.65 millibars), and the minimum at Utopia was 6.8 millibars (compared with 7.89 millibars). About one-fifth of the Martian atmosphere cycles between the polar caps and the atmosphere, moving from the atmosphere to the winter cap, then back to the atmosphere and thence to the winter cap developing on the opposite pole, and so on.

The wind directions at the two landing sites followed repeating patterns each day. At Chryse the wind direction rotated counterclockwise, contrary to what was expected by theory and probably as a result of the local topography of the basin. At Utopia, the wind direction behaved as expected and rotated clockwise each day. Average speeds were 14.5 miles per hour (6.5 meters/sec) at Chryse, and 7.8 miles per hour (3.5 meters/sec) at Utopia. As Mars approached solar conjunction and moved closer to the sun, the wind speeds began to increase. At times the winds gusted as high as 38 miles per hour (17 meters/sec). But these gusts were not strong enough to raise dust clouds in the rarefied Martian atmosphere. The scientists had to wait until after conjunction, in fact until February 1977, before a major dust storm developed. This showed that storms do not have to take place only at the time of southern summer solstice and are more prevalent than previously thought.

Shortly after Mars passed perihelion, the southern summer solstice dust storm did develop (figure 6.16). When Viking 2 photographed a large area of the storm in June 1977, the dust rose to the tops of the Tharsis volcanoes although there were still some clear spots through which the surface could be seen. The high-resolution pictures also showed unexplained linear features in the clouds of dust and, strangely, while high winds obviously were stirring up the dust clouds, just alongside could be seen the great chasms of the Valles Marineris with their floors obscured by mists.

By the end of October 1976, Mars and the four Viking spacecraft, which were still functioning well, reached maximum distance from the earth (235 million miles, 380 million km) on the far side of the sun. Already the sun's corona had begun to affect the radio signal from the Vikings. Orbiter 1 had completed two "walks" around Mars by timing its orbit to cover different longitudes each day instead of being fixed over the landing site, and had mapped the surface of the planet in great detail and inspected the north polar region. Now the spacecraft started to pass behind Mars, out of view of earth, once each orbit. Radio scientists used the interruption of its radio signals to measure the effects of Mars' atmosphere and to measure the radius of the planet more accurately.

Experiments on the surface and in orbit were discontinued on November 5 for the period of solar conjunction, and the spacecraft were placed in hibernation, transmitting only engineering data. The behavior of radio signals from the spacecraft during the period on either side of the solar conjunction was used by radio scientists to check the general theory of relativity. The experiment measured the time required for radio

6.16 This picture, taken June 7, 1977, shows part of Mars' southern hemisphere, which is almost covered by a developing global-scale dust storm. The equator is at the top of the picture and 57 degrees south latitude is at the bottom. The picture extends from 75 to 150 degrees west longitude. Two of the huge Tharsis volcanoes are visible at upper left as dark circular markings. The western part of the giant Valles Marineris canyon system, its floor misted, stretches across the top of the picture just north of the dust clouds. These clouds display considerable linear structure. Dark areas within the clouds are relatively clear spots through which the surface is visible.

signals to travel from Deep Space Network antennas in California and Australia to the spacecraft and then return. These measurements were precise to one ten-millionth of a second. Einstein's theory predicts that the sun's gravity should delay the signals by about 200 millionths of a second when Mars is behind the sun. This occurred on Thanksgiving Day, when the round-trip time of the radio signals was about 2,300 seconds. Results were good and agreed with predictions, thus confirming the general theory of relativity more accurately than any previous test had done.

Commenting on these relativity measurements, Dr. Erwin Shapiro of the Massachusetts Institute of Technology, a member of the radio science team, said, "The result is already under half a percent, and the agreement is very good with general relativity. The best prior test had uncertainties of 1 percent or more. Some of the measurements were made with an uncertainty no larger than the distance from my head to my feet, that is, only a 5-foot error at 200 million miles [322 million km]. That is the highest accuracy of a length measurement ever made in the history of man, and corresponds to an error of only 5 parts in 10 million million."

On December 16 and 17, 1976, the first new commands were transmitted to the two Viking landers, ordering them to gather data at low rates and begin playing them back on December 20. Early in January 1977 the Vikings resumed full activities in what was termed an extended mission to look at events on the surface of Mars and from orbit during an entire Martian year. The orbits of the orbiters were changed, approaching closer to Mars and to the two Martian satellites, Deimos and Phobos. Biology experiments continued until the supplies of nutrients were used up. More photographs were taken and more soil samples analyzed. Seismic events were monitored to clarify two events recorded before conjunction. These two events indicated that Mars is not a cold, dead planet; their characteristics were similar to terrestrial earthquakes and quite unlike lunar seismic events.

The results with both landers had been spectacular. Analyses of the soil at the two sites more than 4,000 miles (6,400 km) apart had produced remarkably similar results, a soil unlike any major soil or rock unit on either the earth or the moon. The soil samples contained silicon, iron, and oxygen as the main constituents, probably in the form of silicon oxide and iron oxide, but with significant amounts of calcium, aluminum, magnesium, and sulfur. The sulfur content was more than 100 times greater than its average in the crust of the earth. Possibly 10 percent of the soil could be of sulfate material, which might be the cement holding the fines together to produce the clods of soil seen and sampled at both sites.

The soil seemed to be best described as an iron-rich clay that is similar to clay formed on earth by the weathering of volcanic rocks and by precipitation in sediments of deep oceans. Dr. Richard Shorthill, of the University of Utah, commented on the Martian surface: "We don't understand the interaction of the footpad [with the soil]. The Martian soil behaves mechanically something like . . . a wet sandy beach, yet it is extremely dry [figure 6.17]. The phenomenon that makes soil stick together on earth is for the most part caused by water. But it may be a different phenomenon that causes the cohesion on Mars, which appears to have only a monoscopic [extremely thin] layer of water. It appears the crust at site 2 is more extensive. When you dig into it, little plates form and the surface domes up. These plates then break off. There were places that looked like dry, crusty material on the surface, yet we penetrated right through them with the soil sampler."

Before conjunction the surface around the landing sites was inspected through more than 1,000 high-quality photographs from the cameras of the landers (see selection of figure 6.18). The surface at both sites was revealed as a rocky, boulder-strewn, reddish-orange desert. Many of the rocks appeared to be vesicular (blistered and pitted) or spongelike, and might have been produced either by volcanic processes or by high-speed impact of meteorites.

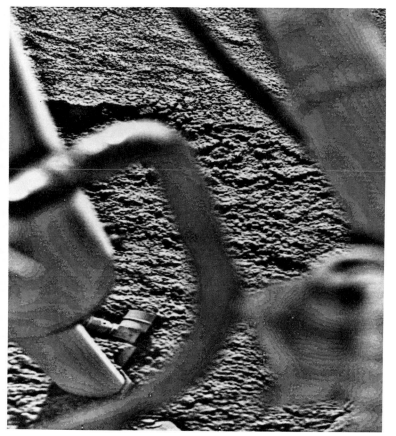

6.17 One of Viking 1's three footpads lies buried beneath a cover of loose Martian soil. The foot sank about 5 inches, and fine-grained material slumped into the depression and over the foot. The cracked nature of the surface near the slumped area, and the small, steep cliff at the left indicate that the material is weakly cohesive.

6.18 Over months of operations on the surface of Mars at two sites, the Viking landers produced a fantastic collection of pictures of the surface under different angles of illumination. A and B show morning and afternoon views of sand dunes at the Viking 1 site in Chryse. C and D show gravel and small drifts in the channel at the Utopia site under direct and low-angle illumination. E, F, and G (also at the Utopia site) show how details in the rocks and the soil are brought out by photographs at different times of the Martian day. H and I show the wealth of detail obtainable with the Viking cameras. These vessiculated rocks are at the Utopia site. Note, however, the angular rock of an entirely different type near the right center of H and the encrusted, platy surface on I.

A.

B.

C.

D.

E.

F.

6.18 G.

H.

I.

A.

B.

6.19 The Utopia site revealed an unusual feature shown in these two pictures: A, two high-resolution scans are put together to show this scene looking northeast to the horizon, which is some 2 miles away. The largest rock, near the center of the picture, is about 2 feet long and 1 foot high. A small channel appears to wind from upper left to lower right, and it is relatively free of rocks. B shows an enlargement of the lower part of the channel. Whether the channel has been formed by running water or represents a debris-filled crack in underlying lava is not known. The wriggly light lines near the right side of this picture intrigued scientists. They selected this area as one of the sampling sites. (Note: The horizon appears to slope because the spacecraft is not resting level on the surface. The horizon is actually quite horizontal at this site.)

6.20 This lander 2 picture from Utopia shows the first clear indication of frost accumulation on the Martian surface. It was taken September 13, 1977. The season is late winter in the northern hemisphere of Mars. Frost appears as a white accumulation around the bottom of rocks, in a trench dug by the lander's sampler arm, and in scattered patches on the surface. It is most probably water frost.

A.

6.21 A prominent feature seen from the lander 1 site in Chryse is the large rock dubbed "Big Joe." *A* and *B* show this rock and its environs under early morning and midday illumination. *C* is a lower-resolution picture taken in August 1977, which shows there has been slippage of the dune material from beneath the rock (see arrows). This movement of the dune material is particularly apparent when *C* is compared with *B*.

B.

Details of the two block-strewn landscapes showed differences and great variety. The Chryse site had exposed bedrock and an undulating surface with a relief (range of elevation) of several meters. The Utopia site did not show any bedrock and was remarkably flat. As seen from lander 1, there was a large field of dunes or drifts of fine-grained material that had been scalloped and eroded by winds to present forms very similar to those seen in terrestrial deserts. No such dunes were visible at the Utopia site of lander 2.

At the site of lander 1 the rocks varied greatly in brightness, shape, and texture. By contrast, those at the site of lander 2 were almost monotonous, and were mainly spongelike. While the second site showed stripping of surface materials by Martian winds, individual rocks at the first site showed more evidence of faceting by wind-borne debris.

The many small, thin, platelike objects seen at the Viking 2 site suggested that the surface was crusted, possibly as a result of water carrying salts to the surface and then evaporating. There were smaller areas of this crust at the first landing site.

The troughs at the landing site in Utopia generally had fewer rocks than elsewhere and had accumulated some drifts of fine material (figure 6.19). The fractured pattern of the surface might have originated when cooling sheets of lava formed polygonal networks. On the other hand, they could have been ice fractures caused by repeated seasonal freezing and thawing of the ground; these are common in Arctic terrains such as at Prudhoe Bay in Alaska.

The lander 2 site was covered with large blocks of a relatively small range of sizes. Viking scientists speculated that these might have been ejected from the large crater Mie, which lies about 125 miles (200 km) east of the site.

One of the most interesting pictures (figure 6.20) taken during the extended mission was obtained during late winter at the Utopia site. Frost appears as a white accumulation around rocks, in a sampling trench, and in patches on the dark surface. The frost apparently formed during the night and partly disappeared during the daytime. The combination of temperature and pressures measured by the lander at the time was inconsistent with a carbon dioxide frost. Scientists thought there was a very good chance that it was water frost.

But apart from frost appearing on the landscape, there were very few changes observed by either lander. One of these was a slippage of fine material at the base of the "Big Joe" rock (figure 6.21). Scientists were disappointed that other changes were not apparent even after large dust storms had been observed from orbit.

c.

Chapter 7

Shades of Van Leeuwenhoek

While the physical properties of the Martian surface were being investigated thoroughly by the instruments carried on the two landers, and the orbiters high overhead continued to survey the planet photographically and in infrared radiation, the search for life began to produce results that were both startling and contradictory.

An experiment referred to as pyrolytic release (Horowitz's experiment) sought to find out whether food was manufactured from carbon dioxide and carbon monoxide as plants do on earth (Figure 7.1). A sample of soil was placed in an incubation chamber in which it was kept dry and without food in a simulated Martian atmosphere. The chamber was sealed, and a small amount of carbon dioxide and carbon monoxide labeled with carbon 14 (an isotope of carbon that behaves chemically like ordinary carbon but can be easily detected by the electrons it emits) was squirted into the test chamber. A xenon arc light was then turned on to shine through a glass window into the chamber, simulating the solar spectrum. Microorganisms would be expected to use the labeled carbon gases and incorporate the carbon 14 atoms into their biological molecules.

If the soil contained photosynthetic organisms, or just organisms that could fix carbon dioxide, results would be expected after 5 days of incubation at about

50 degrees Fahrenheit (10°C). The atmosphere was then blown out through a vent. The chamber was next moved to another position where it sat beneath an organic gas chromatograph and was heated to 248 degrees Fahrenheit (120°C). That temperature was not hot enough to break down organic compounds but would get rid of carbon dioxide and carbon monoxide absorbed in the soil sample. A detector was expected to show a big peak of radioactivity (from the carbon 14) as these absorbed gases were removed from the sample. Once they were cleared, the count of carbon 14 would fall to the background level again.

Next the soil was heated to about 1,160 degrees Fahrenheit (625°C), a temperature high enough to decompose (pyrolyze) any molecules or organic matter into organic gases. Small organic fragments would be trapped in the absorbing column of the gas chromatograph. Then this column was heated to a high temperature to release the organic material and oxidize it back to carbon dioxide, which flowed into the detector. If photosynthesis had taken place within the soil sample, there should be a second peak of radioactivity as the labeled carbon passed into the detector.

This experiment could also be used with the light off to check whether there were any organisms able to use the labeled gases in the dark. Another test could

Shades of Van Leeuwenhoek

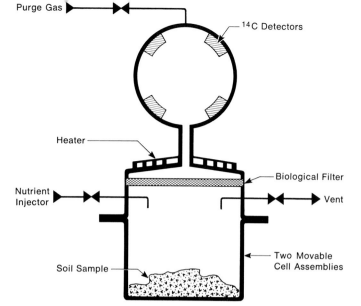

7.1 The pyrolytic-release experiment sought microorganisms that might create organic compounds out of radioactively labeled carbon dioxide and carbon monoxide, with or without simulated sunlight. The soil sample placed in the test chamber was incubated for 5 days. Then the atmosphere was removed from the chamber and the soil was heated to drive off absorbed gases. These gases were detected as a first peak of radioactivity. Next the soil sample was heated to a higher temperature that decomposed (pyrolyzed) any organic compounds. These were trapped in a column and then later released by heating the column. Any trapped molecules were oxidized into carbon dioxide, which was passed through the detector to produce a second peak of radioactivity.

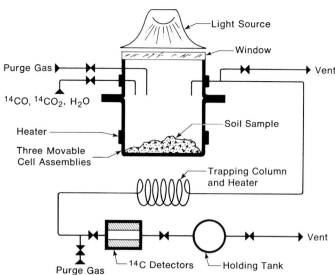

7.2 The labeled-release experiment looked for microorganisms that could metabolize organic (carbon) compounds in a nutrient liquid. These compounds were labeled with radioactive carbon 14. The soil sample was first slightly wetted with nutrient. The gas above it was monitored for radioactivity. Counts in the detector indicated how much carbon 14 was being removed from the liquid nutrient and given off as carbon dioxide or carbon monoxide gas. Later more nutrient was added and the effects were observed. Transfer of radioactivity from the nutrient to the gas in the headspace was thought to be a sign of biological activity.

inject a small amount of water into the atmosphere above the soil, to find out if water in the atmosphere is needed for photosynthesis on Mars.

The second, labeled-release experiment (Levin's experiment), was the simplest (figure 7.2). It supplied a nutrient solution and checked whether some of it was consumed. Any living things in the soil sample would be expected to metabolize one or more of the carbon 14–labeled compounds in the nutrient solution and to release labeled carbon dioxide into the atmosphere. The soil sample was placed in an incubation chamber together with some Martian atmosphere. The chamber was then sealed and kept dark. A very small amount of the labeled nutrient solution was sprayed on the soil, and immediately a carbon 14 detector attached to the incubation

chamber was turned on. The sample was then incubated. If there was life on Mars, and if that life could metabolize the nutrients, then the detector would begin to register a steadily increasing amount of radioactivity.

The third experiment, called the gas-exchange experiment (Oyama's experiment), was the most complex (figure 7.3). It did not use radioactive labels but looked for changes to the environment caused by living things in the soil sample. A soil sample was acquired and sealed in a test chamber in a porous incubation cup raised above the bottom of the chamber. The atmosphere within the test chamber was pure carbon dioxide; for technical reasons it could not be the Martian atmosphere. For the first week the soil was incubated in a humid situation, but was not in contact with any nutrient. Later the nutrient liquid was permitted to rise into contact with the soil.

Shades of Van Leeuwenhoek

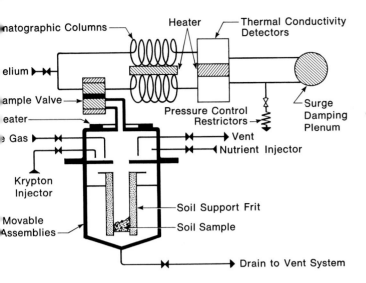

Chromatographic Columns —
Heater
Thermal Conductivity Detectors
Helium
Sample Valve —
Heater —
Gas
Krypton Injector
Movable Assemblies —
Pressure Control Restrictors
Surge Damping Plenum
Vent
Nutrient Injector
Soil Support Frit
Soil Sample
Drain to Vent System

7.3 The gas-exchange experiment looked for evidence of microorganisms exchanging gases with a surrounding atmosphere. The soil sample was placed in a test chamber and exposed to water vapor for a period of incubation. Later the soil was wetted with a nutrient solution. The gas in the headspace was monitored for changes in composition.

The assay system was different from the other two experiments, which depended on carbon 14 detection. This experiment used a gas chromatograph and a thermal conductivity detector to sniff the atmosphere above the soil sample every day during the incubation, to measure concentrations of low-molecular-weight gases such as hydrogen, carbon dioxide, methane, oxygen, and nitrogen. The idea was that one or several of these gases might suddenly begin to appear or disappear as Martian life used the nutrient. This experiment looked at all these gases for changes to whatever the starting amounts were. It was another way of looking for life, because terrestrial life respires, or changes in some way, the gases surrounding it. If change to the gas occurred in the "humid mode," the instrument could be commanded to continue in that mode without introducing any nutrient. If nothing happened in the humid mode, the experiment would proceed to the nutrient mode.

For all three experiments, part of the initial sample was saved in a separate receptacle. If a positive signal was obtained in any experiment, the test could be repeated after the sample was sterilized by heating it to 320 degrees Fahrenheit (160°C) for 3 hours. This would make sure that the positive results were not due to an inorganic chemical reaction of some kind. The heat sterilization would be expected to lead to a negative reaction if the initial positive reaction had derived from the presence of microbes in the Martian soil. If it was a chemical reaction, however, it would

not be affected greatly by the sterilization. The control test of a sterilized sample was an important part of the experiments. Without such a control it would not be possible to say with much certainty whether a positive result indicated the presence of life on Mars.

Each experiment could be done three times. The sequence could be extended, if needed, to allow a fourth analysis as a control if a positive result was not obtained until the third test. The tests were referred to as cycles. Active cycles took place with unsterilized samples, control cycles used sterilized samples of soil.

Gerry Soffen said that, before the flight, everyone worried that the biology experiments would be too complex to work on Mars, or that the results would be so indefinite that scientists could not decide if the instruments were producing positive results. The big surprise was that all the instruments reacted violently to the Martian soil samples. But the idea of these results being an active Martian biology was offset by the depressing fact that the organic chemistry experiment showed no evidence of organics in the soil.

At a science briefing on July 28, Dr. Soffen (figure 7.4) described his elation that morning when he encountered a young engineer racing up the stairs of the Viking Building at JPL shouting, "The PR lamp is on, the PR lamp is on." That engineer's dream had come true when the biology experiment for pyrolytic release (PR) was known to be functioning properly on the distant Chryse plain. The years of work trying to get this instrument to work had paid off. It was now

Shades of Van Leeuwenhoek

7.4 Dr. Gerald Soffen, Viking project scientist, points to the input hoppers of the various experiments that needed soil samples. In front of him is the soil sampler arm (retracted) and scoop with which the samples were gathered from the Martian surface.

more oxygen was released, so whatever released the oxygen at the beginning of the experiment, when the soil was exposed to humidity, had apparently all been used up. This looked more like chemistry than biology. A decrease in carbon dioxide within the test chamber was accounted for by absorption of the gas in the liquid nutrient. "We believe that there is something in the surface, some chemical or physical entity which . . . may, in fact, mimic . . . biological activity."

Dr. Klein continued: "The second point is that in the labeled-release experiment the preliminary data indicate that we are evolving a fairly high level of radioactivity which, to a first approximation, would look very much like a biological signal. That second result must be viewed very, very carefully in order to be certain that we are, in fact, dealing with a biological or nonbiological phenomenon."

At 1:30 A.M. on Friday July 30, two drops of nutrient had been added to the soil in the test chamber of the labeled-release experiment. The nutrient contained

searching for evidence of life on the rock-strewn, pink plain of Mars. This was the first biology experiment to start operating after the soil collector successfully delivered its samples on the eighth day of Viking 1 operations on Chryse Planitia.

A few days later the biology teams were ready to comment on the results. By July 31, Dr. Harold Klein (figure 7.5) was able to announce: "All three instruments are working normally and are doing what they are supposed to do in terms of operations. We're in good shape in terms of the biology experiment." So far, two unique and exciting things had been discovered. "In one of our experiments, the gas-exchange experiment, we believe that we have at least preliminary evidence for a very active surface material." The soil sample in the gas-exchange experiment had released oxygen when placed in a humid atmosphere. Dampening the sample with nutrient did not cause any further change in oxygen release, however. No

7.5 Dr. Harold Klein, head of the team searching for life on Mars, explains the significance of the Viking discoveries to newsmen in the Von Karman auditorium at the Jet Propulsion Laboratory.

Shades of Van Leeuwenhoek

radioactive carbon 14 which, if metabolized by a microorganism, would be released in radioactive gases. The experiment looked for a rise in radioactive counts in the gases in the chamber over a period of time. Dr. Gilbert Levin had developed this experiment. "The data we got back [from Mars] last night," he said, "are consistent with the kinds of responses that we are used to seeing from terrestrial soils."

On the significance of this experiment, Dr. Klein commented: "At present there is no way you can rule out that the data are . . . due to biology. But . . . it's a stronger response than we have seen with fairly rich terrestrial soils. But in view of the unusual chemical properties of the soil, we must emphasize that it is entirely feasible that what we are seeing is the result of the [chemical] oxidation of one or more of the radioactive nutrients that were added in this experiment."

By August 1 the results of the labeled-release experiment were more mysterious. The process, whatever it was, had slowed down. Dr. Klein said, "As we have tested a large number of terrestrial soils, wherever we have had this rapid an initial response from them, we have never seen it [the count rate] slowing down so quickly. Generally, if it goes up fast, it takes a long time to slow down." Levin warned the next day, "We must try every other possibility to explain the responses . . . by chemical means, before absolutely being driven . . . to the conclusion that we can only explain it by [a] living reaction." There was a tentative hypothesis that some unique or rare photochemistry might be operating on Mars.

This experiment produced more surprises when nutrient was injected a second time, after the counting rate triggered by the first nutrient injection had leveled off. Much to our surprise, said Dr. Klein, and to our chagrin, since we cannot explain it yet, after injection of the second batch of nutrient the count rose

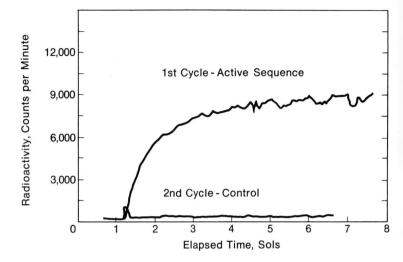

7.6 The labeled-release experiment in its first cycle at the Viking 1 site in Chryse produced these results; if they were obtained on earth, they would be regarded as evidence of a biologically active soil.

about 1,200 counts in the first 8 minutes. "But immediately after that the count began to decay [decline] and dropped to a level about 30 percent lower than it had been. This decay stopped and began to increase again slightly just as the relay link to earth ended" (figure 7.6).

On the other biology experiment, the pyrolytic release experiment, Dr. Horowitz announced on August 3 that pyrolysis (heating to cause decomposition) had taken place. the first peak of radioactivity—from the gases that had been absorbed in the soil—placed the Martian sample in the range of soils that were tested on earth. Four days later Dr. Horowitz was able to report that the second peak in his experiment, which should indicate the amount of radioactive carbon dioxide extracted chemically by the soil from the atmosphere in the test chamber, was 96 counts per minute. The predicted second peak for a terrestrial soil without living organisms was only 15 counts per minute. "There's a possibility that it is biological," he said, but added that there are many other possibilities that have to be excluded first.

Asked if the pyrolytic-release results could have been caused by oxygen-rich molecules of superoxides and peroxides in the Martian soil that might account for the oxygen release in the gas-exchange experiment, Dr. Horowitz replied, "No, I think we can say something very definite about that. What we're seeing is a reduction. Superoxides and peroxides carry out oxidation."

Now everyone waited for the control cycle on this experiment, the testing of another sample that had first been sterilized by heating. Dr. Klein went so far as to say that if organics were found on Mars and if the control came out right, "I think we'd be tempted to say we had found something." It looked very promising for life on Mars. On August 16 the heat sterilization of the sample for the second pyrolytic-release cycle was completed, and the incubation began.

On August 25, Dr. Horowitz said he had proposed three criteria for accepting the results of his experiment as being due to Martian biology:

> 1. The signal must be eliminated by a heat-sterilized control.
> 2. The original observation would have to be reproducible.
> 3. Organic materials would have to be detected in Martian soil.

Reporting on the first criterion, Dr. Horowitz said the control sample had been sterilized by heating to 356 degrees Fahrenheit (180°C) for about 3 hours. This time the second peak of radioactivity was very low—heat treatment did abolish the effect observed in the first cycle. "This eliminates many of the instrument failures we had thought might be causing the positive results of the first cycle."

> We have two major explanations for the first result and one is that there is some unusual Martian surface chemistry going on; chemistry that we haven't seen yet in our laboratory. And the second is that there are organisms in the material at Viking landing site 1. Actually we're talking about chemistries of different orders of complexity. Perhaps the least likely kind of chemistry is living chemistry, and there must be a whole range of possibilities between living organisms and no chemistry at all that we

haven't explored. . . . But we still have to test the reproducibility of this first test, and we still have to find organic matter on Mars before we will be led to announce that we've discovered life on Mars.

> If these counts that we are seeing are due to metabolism, then there must be some place where you find organic material in detectable quantities on the surface. If we don't find organic matter on Mars, it will be very hard to maintain a biological interpretation.

Another sample was scooped up a few inches away from the place where the first one had been taken, early in the morning of sol 8. It was used in the third cycle of the experiment to see if the results of the first cycle could be repeated. But this second sample was acquired in the warmth of Martian midday, so that conditions of the experiments were, unfortunately, not identical. The incubation also was a few degrees higher than the first time. The warmer incubation happened because the biology team tried to keep some of the old soil in reserve until acquisition of a new sample was confirmed; then the old soil could be thrown away. To do this they had to turn off the power for the biology instrument and instruct the instrument to turn itself on again when it received new soil. Inadvertently the coolers of the biology package were not turned on again. It took two and a half days to bring the incubation temperature back to normal.

The first peak of radioactivity was about the same as for the first active cycle, but the second peak was now only 28 counts compared with 96 counts—a big unexplained difference.

These results confused the experimenters. The criterion of being able to repeat the first experiment had not been satisfied, but the reason could be that conditions for the two experiment cycles were not identical. Coupled with the inability of Dr. Biemann's GCMS experiment to find any organics on Mars (the third criterion), biology looked less likely. Everyone hoped that Viking 2 would provide an answer.

Shades of Van Leeuwenhoek

On August 17, a control sample for the labeled-release experiment was sterilized by heating it to 320 degrees Fahrenheit (160°C) for three hours. The results from the control cycle of this experiment could have ruled out the possible presence of biology in the soil if they had been similar to the results of the active cycle. They were not. Indeed, the lack of any rise in the counting rate was consistent with a biological interpretation. Dr. Levin commented: "We waited very patiently through the first days of the control cycle for the arrival of the first data. Right away it is evident that there is a significant difference between the two curves [the control and the active]. Had we seen these two curves on earth—if we had run this experiment in the parking lot at JPL—we would have concluded that life is present in the sample. But regardless of whether the results are biological or chemical, we believe that they are very exciting, very significant; that some highly interesting activity is present on the surface of Mars that I don't believe any of us anticipated at the time Viking was launched. We wish that Dr. Biemann would find a little bit of organics in his next test."

When lunar soil had been put through the same tests, unsterilized and sterilized, the results had been quite unlike those obtained from the Martian soil. The Martian soil results resembled most closely tests of Antarctic soils. The experimenters concluded that a good control cycle had been obtained. But what this meant in terms of life on Mars was still unclear.

Starting at sol 39, the labeled-release experiment began a 66-day incubation to see if there was any exponential rise in counts after the second injection of nutrient, as was suspected during the first cycle. An exponential rise would be a good indication of a Martian microbe thriving on the nutrient. This second active cycle produced results similar to the first one. Said Dr. Levin on September 4. "Comparison of first and second cycles [active and control] would on earth

have been regarded as biology. The purpose on cycle 3 was the same as for pyrolytic release—to try to duplicate the first experiment, but also to add some incubation time. The experiment repeated completely. We got the vigorous response that we received the first time, and the shape of that response is essentially the same." Although the temperature of the sample was about 16 degrees Fahrenheit (9°C) warmer than the first sample as it was placed in the incubation cell because it had been taken later in the Martian day, "We would say . . . an excellent repeat of the experiment at this point."

On August 23, Dr. Klaus Biemann reported on a search for organics in a sample obtained on sol 31 from an area that he referred to as "rocky flats." He said, "We found, as last time, copious amounts of water and little or no organic material. Organically speaking, both samples were very clean material. The detection limit for organics in this experiment on Mars was about 1,000 times below the normal [level of] organics found in meteorites."

Commenting on the limits of the organic-analysis instrument's ability to detect organics, Dr. Biemann pointed out that the instrument could detect 1 part of organic material in a billion. A typical microorganism weighs 0.000000000001 (or 1×10^{-12}) grams. Of that amount 90 percent is water and 10 percent is dry weight, i.e., organic material, of which only a part will be detectable by the instrument. To register in the instrument the soil sample would have to contain 1 million microorganisms, dead or alive. The small sample acquired by Viking, only one-tenth of a cubic centimeter of soil, was not likely to contain that many living microbes, so the instrument was not expected to detect life.

If there were some living organisms on Mars, however, the experimenters assumed that there would be many more dead organisms in the soil, and it was the organics from these dead organisms that they hoped to find. Since the experiment could detect organic compounds present in an amount 1,000 to 10,000 times smaller than the amount found in a typical carbonaceous chondrite meteorite and yet it had found none, this lack limited the amount of organics that could have been added to the Martian soil by incoming meteorites. Such compounds must be broken down in some way, possibly by ultraviolet radiation. If there is a Martian biology, continued Dr. Biemann, "one would have to assume certain Martian microorganisms are efficient scavengers or cannibals, and very efficiently utilize dead microorganisms." Dr. Mike McElroy added that

> whether it's biology or chemistry, the Martian soil obviously likes to destroy reduced carbon [i.e., organic materials]; reduced compounds are oxidized [used] fast. The real puzzle is that the third experiment [pyrolytic release] seems to say that the Martian soil likes to make [reduced carbon] as well. Now that is difficult to do with just chemistry. Which is not to say that we are observing biology, but we are seeing a very, very strange chemistry. The organic analysis . . . is not very positive or negative on this. There are good reasons to believe that Mars doesn't like to hold reduced carbon, and so all dead organisms would not be present in the soil.

> Their carbon would be back in the atmosphere. We are left with an extremely puzzling thing. I think the fact that the soil releases oxygen is one of the most exciting results of Viking. Look at the past history of Mars. Abundant amounts of water wetted the soil. At that time the planet's surface would have released a vast amount of oxygen. If oxygen is emitted it would be converted to ozone, which would cut off the ultraviolet radiation and stop photo-oxidation of the surface. How does the oxygen get back into the surface?

The gas-exchange experiment had started its second cycle on August 18. All the liquid nutrient was drained from the test chamber, the gas was purged from the chamber, then a fresh amount of nutrient was introduced and the same soil sample continued to incubate. After the flushing of the atmosphere and the replacement of the nutrient solution, no further oxygen was released. This result was as expected because this chemical quality of the soil had apparently been used up. But a strange effect was that the carbon dioxide concentration increased in the atmosphere above the incubating soil (figure 7.7). The increase was large, and its cause was not clear. Speculation ranged from a vestige of the oxygen-releasing activity to biological activity.

The biology experiments of Viking 2 also produced puzzling results. By September 14, Viking 2 had acquired soil samples and the experiments began. The gas-exchange experiment (changes to the environment) soon returned interesting data. Gas analyses made before the soil sample was inserted into the test chamber were consistent with the mass spectrometric analyses of the atmosphere. With soil in the chamber, the analyses showed the same gases plus a

7.7 The gas-exchange experiment at the lander 1 site produced these data during its incubation period of 89 sols; the vertical dotted lines represent the beginning of fresh incubation cycles using the same soil sample but with renewed nutrient.

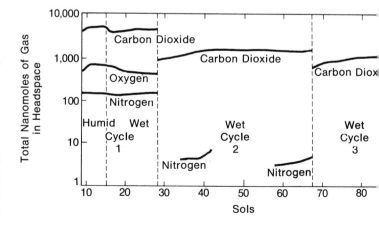

little extra carbon dioxide. After the soil was humidified with water vapor, oxygen was released, but in smaller quantity than from the soil sample of Viking 1. This result suggested that the soil at the Viking 2 site might contain considerably less of the oxidizing substances than the soil at the Viking 1 site.

Dr. Vance Oyama commented on the gas-exchange experiment at a briefing on September 23; he was finding significant differences between results at the Chryse and Utopia sites. Samples at Chryse produced much more oxygen than those at Utopia, but for carbon dioxide the results were almost the same. "The oxygen only came out when we humidified the soil and is a result of what I think is chemical reaction between water and such a thing as superperoxide in the soil material. This oxygen production is a result of water displacement of very labile oxygen groups in the soil itself." Such groups would be readily released from the soil when water was added to it.

The experimenters wanted to see what would happen at Utopia—whether oxygen was released because of a humid atmosphere or because of warming when placed in the test chamber, which was warmer than the Martian surface. They found that it was not warming of the soil but humidity that released the oxygen, even though at Utopia the amount released was only about 20 percent of what it was at Chryse.

The differences in oxygen release could be simply a matter of the higher vapor concentration in the atmosphere over Utopia compared to Chryse. The Chryse soil would be less exposed to water vapor and therefore could hold more oxygen for release when wetted in the test chamber. At Utopia, however, atmospheric moisture would release oxygen under the normal Martian conditions.

Asked to evaluate the possibilities of life on Mars at this stage of the search, Dr. Oyama said:

We have a very pervasive chemistry that reaches down to the pit of this hole we've dug, some couple of inches below the surface. The chemistry that we see is a highly oxidizing one. It is not oxygen that has been desorbed [released from the soil] at the lower temperature of Mars, but it is oxygen that is coming out of a chemical reaction which has potency for oxidizing organic species. This may account for the GCMS lack of organic detection. The thing which is interesting about this phenomenon is that at site B [Utopia] it is less than at site A [Chryse]. It is less to the extent that it is inversely proportional to the amount of water vapor that is in the atmosphere. Where there is more water vapor in the atmosphere, there is less oxygen produced from the soil.

However, there is as much combustion of carbon dioxide, as is demonstrated in the labeled-release experiment, which would suggest that the oxidizing potential has changed its form up north. But it is still there. On a global scale it looks pretty bleak [for life] as far as these landing sites are concerned. I would suggest that the best place to go to search for life might be further south. . . .

If we wash out the carbon dioxide activity, then I am certain that everything we have seen before is pure chemistry. If the carbon dioxide values . . . persist, and especially if [they are] accompanied by other gas changes, then we ascribe that to biological activity.

For Dr. Levin's labeled-release experiment, the data received from lander 2 showed a strong release of radioactive gas when nutrient was added to the soil. The results imposed a new constraint on chemical theories to account for the response seen. Dr. Levin commented on results so far from his experiment:

You may recall [that at Chryse] we had a very swift onset of evolution of radioactive gas when radioactive organic compounds [in the nutrient solution] were placed on a small portion of the soil. When a second quantity of the radioactive compounds was injected onto the soil, we saw that some of the radioactive gas went into solution, disappeared from the detector, and we followed this for 7 more days. At the end of that time we began to see that more gas was evolved, the downward slope reversed and started going upward. We even had some intriguing evidence that it was rising slowly but at an exponential rate.

Shades of Van Leeuwenhoek

Therefore, on cycle 3, at the same landing site, the one that is currently operating, we endeavored to repeat this experiment but to extend the period after the second injection to see if, indeed, we got a rise that became exponential. This would be convincing evidence for the existence of a living process.

On cycle 3 we got a very strong response, very similar to cycle 1. We ran it a little more than twice as long before we injected a second quantity of [nutrient] medium, which then, just as in the first case, produced a diminution in the amount of gas available in the detector. This [the count rate curve] then did flatten out. In the last few days the slope reversed and is positive, indicating that gas is being evolved again. However, it is too early to say whether this evolution is of an exponential rate, or is merely a straight line. A straight line would not be nearly so convincing, since it would not indicate that growth was taking place.

The drop-off in counts after the second injection of nutrient liquid in the experiment was assumed to be caused by absorption of the radioactively labeled gas back into the liquid. Dr. Levin further commented:

> The experiments that are being played out now have been 16 years in planning. It's a very difficult question. If we were getting these results [of the labeled-release experiment] any place on earth, we'd say there's life there, and we would be certain of it. The results are remarkably consistent. The tests and the controls all do the things they are supposed to do, except we are on Mars, and the question is a little more momentous. I want to stress at this time, the chemical theory is every bit as far-fetched as the biological theory [for the results obtained]. I don't know whether there is life on Mars. I feel certain at this point we do not have any evidence to rule it out.

Excellent repeatability was achieved on the third cycle. "So, in essence," said Dr. Levin, "we now have three positive active responses [2 at Chryse and 1 at Utopia] from the labeled-release experiment which are all in fairly good agreement, and we have one control [at Chryse], which shows a negative response. Further, we now have two commandable injections, or second additions, of nutrient to active cycles, and these are in agreement with each other."

There were several theories proposed to explain these results. First, some of the organic nutrients might have been catalytically changed by one or more rare elements in the Martian soil that would act as a catalyst in the presence of, say, formic acid (which was a constituent of the nutrient) and would liberate radioactive carbon dioxide. The control cycle largely dismissed that possibility, because there was no response after heating to 320 degrees Fahrenheit (160°C). Such heating would not have been expected to change a chemical response of this type.

The second chemical theory, the most prevalent one, was that a strong oxidant, such as a peroxide, a superoxide, or an ozonide, was present in the Martian soil. This was responsible for the decomposition (decarboxylation) of one or more of the organic nutrients, probably formic acid, giving off radioactive carbon dioxide. Again, the control reduced such a possibility.

Dr. Horowitz reported next that Viking 2's first pyrolytic-release experiment (assimilation by the soil sample of radioactively labeled carbon dioxide) had been made without simulated solar radiation in the test cell. The artificial sunlight had been left off to avoid having the lamp heat up the chamber by 4 or 5 degrees. It was also done to test a theory first proposed some 15 years before the Viking mission, that life on Mars might not depend on the kind of photosynthesis that is common on earth. Organisms on Mars, he said, might have developed ways to use carbon monoxide produced in the Martian atmosphere by solar radiation. Such organisms might convert carbon monoxide to carbon dioxide and thereby obtain energy. Some organisms, indeed, do this on earth. Similarly, Martian organisms might live near the surface of the soil, where they could get carbon monoxide from the air and yet be protected from the solar ultraviolet radiation by a thin layer of soil.

The temperature within the test cell ran close to 55.4 degrees Fahrenheit (13°C) for the incubation in the dark without any moisture being added to the soil sample. The first radioactive peak reached 7,133 counts per minute. This peak was produced as labeled carbon dioxide and carbon monoxide gases were driven off from the soil by heating it above incubation temperature. This was the same kind of first peak that was obtained at the Viking 1 site. The second peak, which showed the amount of carbon chemically removed from the atmosphere by the soil and turned into organic compounds, was higher than would be expected from a sterile soil, 21 compared with 15. This was a marginally positive result, said Dr. Horowitz. By terrestrial standards it was equivalent to the biological activity of only 50 *Escherichia coli* bacteria in the soil sample, very many times less than would be expected in a typical sample of a terrestrial soil.

Gentry Lee, the director of science analysis and mission planning for Viking, commented:

> I think it is very important that all of us understand where we are, and where we are not, in the unraveling of the puzzles that Mars has presented to us. I would like to state it very simply that on the basis of the data that we had received up until two or three weeks ago, there were two working hypotheses to explain the data. One of these is that everything we were seeing could be explained by biology. The other was that it is chemistry alone.
>
> Now neither of these hypotheses is readily consistent with all of the data that we have, and just to belabor the point, the biology hypothesis, in order to be totally consis-

tent, has to have some way to explain that no organic compounds have been found. On the other hand, the chemical hypothesis still needs a good explanation of Norm Horowitz's pyrolytic-release data [see figure 7.8] in order to be completely consistent with the evidence we have.

> We have today two competing ideas. . . . I think it is fair to say that the preponderance of scientific opinion is that most of what we have seen can be more easily explained by chemistry; however, it is by no means that clean cut. We need to do many more experiments on Viking and on the ground [i.e., in laboratories on earth] to unravel this puzzle.
>
> The biological hypothesis has only the question of "where are the organics?" yet to answer. Quantitatively, one can have small amounts of living organisms which Norm [Horowitz] may be finding, and insufficient numbers of dead organisms, debris, so that Klaus [Biemann] could detect them. I think that this is a point that has been missed in some of the oversimplifications that have been describing the experiments.
>
> There does exist a plausible hypothesis that says there cannot be enough organics for Klaus to detect them, and still be biology of the quantitative nature that Norm saw in his first experiment.

Asked if one had to conclude that there was no Martian life, Dr. Gerry Soffen answered, "Dr. Biemann's experiment has examined two samples that are less than a thimble-full of soil from Mars, and we're going to make deductions as broad as whether life has ever evolved on that planet based upon that amount of soil? That doesn't sound like a very heroic thing to

Data From the Pyrolytic-release Experiment

Experiment No.	Locale	Conditions	Incubation Temperature	Peak 1 cpm	Peak 2 cpm
1	Chryse	Light, dry, active	$17 \pm 1°C$	7421 ± 59	96 ± 1.15
2	Chryse	Light, dry, control	$15 \pm 1°C$	7649 ± 60	15 ± 1.29
3	Chryse	Light, dry, active	$13°–26°C$	6713 ± 58	27 ± 0.98
4	Utopia	Dark, dry, active	$13° (12°–18°C)$	7133 ± 58	23 ± 1.7
5	Utopia	Light, wet, active	$18° \pm 1.5°C$	$12,523 \pm 76$	2.8 ± 0.92

7.8 The pyrolytic-release experiment produced tantalizing and confusing results at both sites. Some could be regarded as evidence of Martian life if they had been obtained on earth. Others showed no such evidence; possibly some instrument errors crept into the later tests.

do, considering how many years we have put into it. We are so used to seeing our own terrestrial models that perhaps we're looking too hard in familiar areas and not in unfamiliar areas. We're just not used to data like this. We're not used to seeing biology data like this, or even not so much used to seeing organic analysis data like this."

Tom Young said, "I think the question, . . . Is there life on Mars? has not been answered. I don't think the data fully support either biology or chemistry."

An Overview

By the time that Mars passed behind the sun in November 1976, the three experiments seeking evidence of microbial life on Mars had completed several cycles at each landing site. The experiments revealed unexpected chemical activity in the surface material, which in some ways resembled biology but in others might be explained by a complex chemistry.

The first sample for the labeled-release experiment produced a definite positive result (figure 7.6), except that the addition of further nutrient caused a 30 percent drop in the radioactive counting rate. One explanation for this effect was that carbon dioxide in the headspace of the chamber entered into solution in the water of the injected nutrient. A sterilized control sample of Martian soil showed no release of radioactive gas and enhanced the biological interpretation of the first positive test. A third cycle incubated the sample for a much longer period to see if there was any exponential rise in the counting rate after the second injection of nutrient, a type of increase that would be expected if a growing organism were present. There was no such increase.

The first labeled-release sample at the lander 2 site produced much the same positive results as the first sample at the Chryse site, except for a higher counting rate. Since the incubation temperature was lower

at Utopia, the higher counting rate seemed strange. Exactly the reverse would have been expected of either a biological or a chemical activity.

The control sample of Utopia was sterilized at a lower temperature of 122 degrees Fahrenheit (50°C) to check the theory that a living system would be affected more than a chemical system by this temperature change from the 320 degrees Fahrenheit (160°C) sterilization cycle conducted earlier at Chryse. Another surprise! The counting rate inexplicably wiggled up and down randomly for a while and became a daily pattern after several days but close to that of a sterile sample.

The gas-exchange experiment produced results antagonistic to Martian life (figure 7.7). This experiment sought evidence of oxidation and reduction taking place in a soil sample at the same time when it was exposed to a nutrient medium. Such processes occurring simultaneously would be accepted as evidence of biology. The first cycle of the experiment at Chryse incubated the sample for 7 days. Twice the predicted amount of nitrogen was released, and there was a large, unexpected release of oxygen. There was also a characteristic drop in the amount of carbon dioxide after wetting of the sample by nutrient solution. The results could all be explained by chemistry without needing any biological activity.

At Utopia, oxygen was released in the humid condition, but the amount was less than at Chryse, perhaps because Utopia Planitia is exposed to more atmospheric water than is Chryse.

In the release of carbon dioxide the experiment performed as expected if the surface of Mars were a dry surface material absorbing gases for a long time. But the release of oxygen was unexpected. The sudden onset of this release made it unlikely to be a result of a Martian microbe population in the soil waiting to produce oxygen and then suddenly stopping doing so.

The pyrolytic-release experiment produced results that also mimicked biology (figure 7.8), high first peaks and low second peaks that exceeded the background level and were higher than the second peak in sterilized control cycles. However, the sample from Utopia produced lower second peaks than samples tested at Chryse. One possible explanation was that some kind of chemical process took place within the soil that led to organic synthesis, but that this process was not necessarily a biological one. However, the results did not rule out the possibility that a biological process was acting in the Martian soil.

The absence of any organic molecules in the Martian soil argued against the existence of a biological system that acted in the same way as biology acts on earth, with many dead bodies of plants and animals lying around in the terrestrial soil. But terrestrial microbes are great scavengers, and this may also be true on Mars. The pyrolytic-release (carbon assimilation) results are the most difficult to account for without biology, but the fact that adding water vapor slowed down the process is not easily explained.

Nonbiological explanations of what was observed by Viking on Mars are undoubtedly going to be very complex. Dr. Klein pointed out how some questions of terrestrial biology took many decades to resolve:

> Let me remind you that in 1674 Anthony van Leeuwenhoek, who invented the microscope, first wrote about those marvelous little "animalcules," he called them, which appeared soon after he boiled up some hay in some water and let it sit. Within a few days of leaving this infusion of hay sitting around in his laboratory he saw, in his microscope, these little creatures swimming around violently within the broth.

> Now these findings began a long series of debates in the scientific community, and these debates tried to explain where these little things came from, what they were the

result of. There were two schools of thought that developed in this battle that started in 1674. One school felt that this was chemistry; namely that in the hay and water there was some strange chemistry which when put together and heated up started a chemical process which ultimately resulted in these little animalcules; and there were many scientists who believed that.

> There were other scientists who began to question and began to think that the origin of these things was not in the chemistry but that they must have come from another living thing, from a parent animalcule; and so, the biologists of the day set out to do experiments to prove that these animalcules came from other living things and not from the chemistry of the hay and the water. Every time the biologists did the experiments, the chemists would shoot them down. They would find some loophole that disproved the experiment. For 200 years scientists battled over the question of whether the process that they were talking about was chemistry or biology. It was finally settled in the late 1800s.

> Louis Pasteur, who worked in the mid-1800s, lived in the wine country of France. The local understanding of the process of fermentation was that this was chemistry; that there was something inherent in the molecules of sugar which in time caused them to break down and turn into alcohol and carbon dioxide. It took 50 years of experimentation on the part of Louis Pasteur and his graduate students and associates to disprove that and show that these yeasts that everybody saw in the fermenting grape juice, which were considered to be a chemical byproduct inherent in the sugar molecule, were the cause, and that fermentation was a biological process. It was a metabolic process due to living things.

> Let's not separate the biology from chemistry. The thing that's going on even in organisms is chemistry, and it's only a question of whether the chemistry we're looking at is inside of a living system, or whether it's outside or exclusive of a living system.

Joshua Lederberg added to these thoughts:

> In trying to develop a perspective about what's going on here, I was rather quickly reminded of Otto Barberg and his experiments of 50 years ago in trying to understand biological oxidation. He was the German biochemist who laid much of the groundwork for our present understanding of how oxidation is done within the cell. He

spent a considerable period of his scientific life studying simulations of this phenomenon, and what he used was charcoal which was impregnated with iron salts. He was able to show that these inorganic preparations that have no living things in them—they have been heated to a very high temperature beforehand—were able to conduct oxidations of amino acids in the presence of oxygen, liberating carbon dioxide. Not bad models for what we are seeing on Mars. He was using them as models for what goes on in the cell.

Life is an evolving system. If you have objects of such a degree of organization and complexity that they can continue to evolve into higher complexity, then the answer is unambiguous. The borderline cases are the hardest. What is plain is that Mars is seething with chemical activity on the surface and to a degree that one has to say is surprising. To say that these results could be explained by nonorganic phenomena is not to exclude that the chemistry that we're talking about is going to be inside cells. What is startling, surprising, and provocative is how chemically active that soil surface is.

The question of life on Mars is very significant to our understanding of the evolution of life on any planet. Viking made several important discoveries that suggest that Mars may have had conditions conducive to life. Mars has nitrogen in its atmosphere, a necessity for terrestrial-type life. Viking confirmed the presence of channels all over the planet, suggesting large flows of water and possibly abundant water and rainfall in the past.

Viking's measurements of the composition of the atmosphere suggest that the total atmospheric pressure of Mars was probably much higher in the past than it is today. The surface materials of Mars might be explained by the presence of clays which can act as catalysts to make amino acids into the proteins of living things.

The only instrument on Viking that could really tell if there was life on Mars was the imaging system, and the cameras showed nothing attributable to life. There were no signs of large-scale life—no trees, bushes, or grasses. However, Viking looked at only a minute part of the total surface area of Mars, and for a relatively short period of time. Even on earth there are life forms such as the "water bear" which survive dehydrated for as long as 100 years and only spring into life when they have access to water again. Martian life might not be ubiquitous as life is on earth. It might be confined to specific regions which Viking did not sample.

Despite elaborate attempts to reproduce photosynthesis nonbiologically on earth, chemists have not been very successful. Yet on Mars something in the soil is accomplishing some kind of photoreduction, synthesizing carbon compounds, which is an important discovery. The surface material of Mars cannot of itself reduce carbon, as was discovered in the pyrolytic-release experiment. Where is the substance that provides the electrons or the hydrogen to reduce the carbon? If it is water, is it bringing about the reduction by inorganic or biologic mechanisms? If it turns out that there is life on Mars, one of the great secrets that we would hope to learn is how it manages with such a small amount of water.

The pyrolytic-release results—the creation of organic compounds from carbon dioxide in the atmosphere—are the most difficult to explain nonbiologically. But the compounds are in extremely small quantities compared with those in typical terrestrial tests. The addition of water vapor produced puzzling results. If it is biology, why should water stop it? Organisms may be good scavengers of organic material and may metabolize quickly and die because of the excess of water. The absence of organic meteoritic material on Mars may mean that there is a biological scavenging system that consumes it. Meteoritic organics on the moon could have been detected with the Viking instrument.

The fact that we did not find organics in the soil on Mars does go against life, however, except that a scavenging quality could account for it; but Viking cannot supply data to confirm or deny such a possibility. Viking seeks something akin to nonscavenging terrestrial life, not a scavenging Martian life. There is no easy way to show that the results are not due to biology. If chemists on earth cannot duplicate what Viking found on Mars, we will be able to accept those results as being due to Martian biology. This may take many years of experiments in terrestrial laboratories.

Nothing from the Viking's experiments, however, indicates clearly that life exists on Mars. There is certainly a complex chemistry, but it may not be the chemistry of living things. The fact that no organics were detected at the two landing sites drives exobiologists toward a nonbiological explanation; this may be a trap because it assumes that Martian life should be like terrestrial life. And if we are able to prove that there is no life at either the Chryse or the Utopia site, we are still left asking, What about the rest of the planet? The polar regions? The oases? The calderas? The chaotic terrain? Even inside Martian rocks? Very recently, scientists have discovered bacteria thriving inside rocks of Antarctica, where they have established a protective ecological niche.

The question of life on Mars seems destined for extended debate, reminiscent of the era of van Leeuwenhoek, that might only be resolved by a vehicle that can move about on the Martian surface to sample many places, or by the return of samples from Mars for exhaustive examination in laboratories of earth, or even a manned expedition to the Red Planet.

While Viking has not answered our major question, its confusing results have certainly whetted our appetites to know more about this chemically active planet. The riddles of Mars have been rewritten in a new language.

Shades of Van Leeuwenhoek

A.

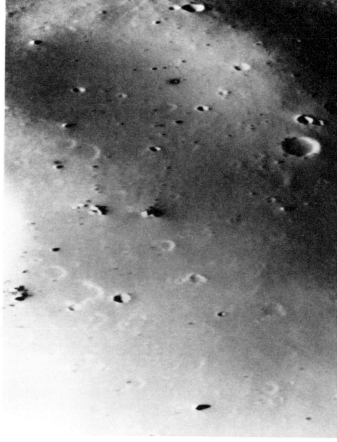

B.

8.1 Viking produced excellent new pictures of the Martian satellites, showing much finer detail on the surfaces of these tiny worlds. *A,* this picture of Deimos was obtained by orbiter 1 from a range of 2,050 miles (3,300 km) early in September 1976. The lighted portion, representing about half of the face of the satellite toward the camera, measures 5 by 7.5 miles (8 by 12 km). The two largest craters on this view measure 0.6 miles (1 km) and 0.8 miles (1.3 km) in diameter. The smallest are only 330 feet (100 m) across. Deimos is the outer, smaller satellite of Mars.

B, this close-up of Deimos was obtained by orbiter 2 on October 15, 1977 from a distance of only 30 miles (50 km). It shows features as small as 10 feet (3 m) across. A layer of dust appears to cover craters that are less than 165 feet (50 m) across, making the surface look smooth. Ghosts of these craters can be seen through the dust layer. The big blocks on the surface are the size of houses.

Chapter 8

Icefields Colored Red

After the orbiters had completed landing-site surveys and the landers had arrived safely on the surface, the orbiters were available for surveying Mars in greater detail than had been possible from the Mariner 9 spacecraft in 1971.

To do this, the orbiters were "unlocked" from their orbits, which were synchronous with the rotation of Mars and brought them to periapsis (closest approach to Mars) over the landing sites each Martian day. The propulsion unit fired and the orbit changed slightly so that the periapsis point moved slowly around the planet. Project personnel referred to this as "walking" the spacecraft around Mars. As they walked, the orbiters took high-resolution pictures of vast areas of the Red Planet, revealing a wealth of detail never seen before. They also extended planetwide their infrared observations of temperature and water vapor concentrations.

Orbiter 2 was able to obtain unprecedented pictures of the north polar regions. Close approaches were also made to Phobos and Deimos, providing fascinating details of these small moons.

When I saw the first close-up photograph of Phobos returned by Viking, revealing many inexplicable features of that tiny satellite's surface, I was reminded of

the first time I ever saw the satellites of Mars. Just after the close opposition of 1971, I visited JPL's Table Mountain Observatory in the San Gabriel Mountains of Southern California. Mars was a brilliant red star in the southern sky in the early hours of the morning. In the 24-inch (61-cm) telescope I could see the usual fuzzy dark markings on the red disc of the planet, and the bright spot of the south polar cap. By turning my eyes away from the bright disc of Mars and looking toward the side of the field of view—a trick well known to amateur astronomers when seeking faint objects—I detected the two points of light that were Deimos and Phobos. Close to Mars and almost hidden in the aura of light from the planet, they winked in and out of view in a tantalizing fashion.

Through the eyes of a Viking orbiter 5 years later I was able to see these tiny specks of light as unusual miniature worlds, heavily pock-marked with craters (figure 8.1).

Using a camera-slewing technique to compensate for smear motion as it rushed closely by Phobos, Viking 2 produced the first close-up of this satellite in September 1976. The surface details were surprising. There were grooves across Phobos' surface that were approximately parallel to each other but tilted about

Icefields Colored Red

30 degrees in relation to the equator. They made the satellite look as though it was carved from a layered chunk of rock. There were also unexpected chains of very small craters; some of them exhibited the familiar herringbone pattern seen in secondary crater fields on the lunar plains, made by rocks thrown up by primary impacts and pulled back by the lunar gravity.

Since Phobos' gravity would be too low to pull debris from impact craters back to the surface with enough velocity to make these striations and secondary crater chains, other solutions had to be sought. One possible explanation was that Phobos was hit by debris moving in orbit about Mars; striations, for example, might be caused by Phobos overtaking a cloud of debris moving in a similar orbit, so that the relative velocities would be almost equal and the pieces of debris would virtually roll across the surface of the satellite. A further speculation was that the gravity of Mars is pulling the satellite apart and it will soon become a ring system, possibly within 100 million years. The herringbone patterns might have been caused by debris that was flung off from impacts and met Phobos again in a later orbit around Mars.

A detailed mosaic of Phobos (figure 8.2) was obtained early in 1977, during the extended mission, and a picture of the satellite against the background of Mars was also obtained (figure 8.3).

If Mars had presented mysteries, its two satellites had been even more puzzling. Their small size and proximity to the planet made them unique in the solar system, and very hard to see. A series of observations of Phobos in the late 1960s suggested that the satellite might be revolving in its orbit in a period that was slowly changing. Such changes could be explained, said the Russian astronomer Iosif Shkovskiy, by Phobos being a hollow world of extremely low density.

The announcement brought speculations that Phobos might be an artificial satellite of Mars. Later observations showed that the orbital motion of Phobos was varying only slightly, and the need for a hollow world no longer existed. However, in the Viking mission, when the orbiter approached Phobos closely, the tracking data again raised interesting speculation about its true nature.

These data revealed that Phobos is a very light world, much less dense than would be expected if it were made of the same material as carbonaceous chondrites. The mystery deepened. Also, it now seems clear that Phobos is speeding up in its orbit as it falls toward Mars under the influence of tidal forces.

Counts of the number of craters of various sizes on Phobos and Deimos indicate that the surfaces of both are completely saturated by craters, like the highland areas of the moon. Even the inside floors of large craters are saturated. "Saturation" occurs when younger craters have completely obliterated older craters. There are no undisturbed areas of the original surface. From this finding it has been concluded that the surfaces of these satellites must be at least 2 billion years old. Also, from photographic and infrared observations from the spacecraft and polarization measurements from earth, it seems clear that the surfaces are finely powdered; no solid rock is exposed.

The orbiter's cameras took color pictures of the two satellites, which are unusually dark objects. They reflect only about 6 percent of the sunlight falling on them (the earth's moon reflects about 11 percent). They are also quite colorless gray objects. The high-resolution pictures obtained by Viking showed that although there are bright and dark areas on the surfaces, these are all a drab gray. This suggests that the surfaces are covered by a very opaque layer of material which is so dark it may be carbon.

Icefields Colored Red

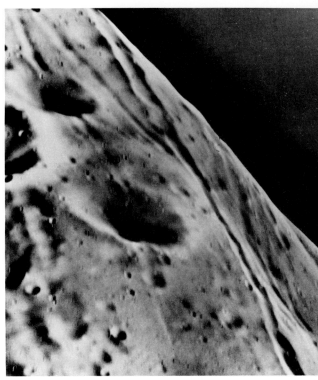

A.

B.

8.2 On February 18, 1977, Viking orbiter 1 flew by within 300 miles (480 km) of Phobos and obtained this high-resolution mosaic (A). The satellite is about three-fourths illuminated and measures 13 miles (21 km) across and 11.8 miles (19 km) from top to bottom. The south pole is within the large crater, Hall, at the bottom center where the pictures overlap. This crater is 3 miles (5 km) in diameter. The pictures reveal details as small as 65 feet (20 m) across. Two days later the Orbiter obtained this close-up, B, from a distance of 75 miles (120 km). The picture covers an area 1.86 by 2.17 miles (3 by 3.5 km), revealing surface details as small as 50 feet (15 m). The area shown is in the northern hemisphere of Phobos and shows some of the characteristic striations on this tiny world. The small craters appear elongated because of smear produced by the high speed of the spacecraft relative to the satellite.

8.3 This picture from the Viking 1 orbiter shows Phobos silhouetted against the surface of Mars. The spacecraft was 8,500 miles (13,700 km) above the surface of Margaritifer Sinus, and 4,160 miles (6,700 km) from Phobos. This moon is one of the darkest objects in the solar system, four times darker than the average surface of Mars. That is why Phobos appears black on the picture, which was exposed for the Martian surface brightness.

The walks of the orbiters around the planet also produced much detailed information about the temperature of the Martian atmosphere and the concentrations of water vapor. Temperatures were measured by infrared thermal mapping. Along latitude 48 degrees north, the temperature varies daily by at least 27 degrees Fahrenheit (15°C), with the highest temperature, −126 degrees Fahrenheit (−88°C), occuring at about 2:15 P.M. local time. This pattern implies significant absorption of sunlight in the lower atmosphere, probably by dust particles.

Temperatures measured elsewhere showed large daily variations at high altitudes, for example, the southern part of the Tharsis plateau. The same was true of the summits of the large volcanoes. For instance, the temperature at the top of Arsia Mons in the early afternoon was twice the nighttime minimum.

Just before solar conjunction Dr. Hugh Keiffer, the leader of the thermal mapping team, discussed some implications of the temperatures measured for the polar caps of Mars:

> The cap which grows and shrinks over a good fraction of the polar regions each year is carbon dioxide. Its temperature has been measured as 150 Kelvin* [−189 degrees Fahrenheit, −123°C], as one would expect for carbon dioxide at the Martian surface pressure. But the question was that of the nature of the residual polar cap. If this was carbon dioxide, we had the possibility of having an atmosphere which could, at least in theory, undergo climatic instabilities and [lead to] at least two different types [of climate] . . . the dry, low-atmospheric-pressure Mars we see at the moment, and the potentially wet, massive-atmosphere Mars which would occur if something triggered sublimation [vaporization] of the polar deposits. If the poles were carbon dioxide, they would tend to buffer the atmospheric pressure through the year.

* The Kelvin scale begins at absolute zero (−473 degrees Fahrenheit), and one Kelvin equals one degree Celsius; 273 Kelvin is 0 degrees Celsius, and 32 degrees Fahrenheit.

The meteorology results had earlier shown, however, that atmospheric pressure fell as carbon dioxide froze out of the atmosphere at the caps. And now the temperatures measured from orbit were too high for carbon dioxide residual caps: "The dark regions [of the residual polar cap] have temperatures on the order of 235 to 240 Kelvin [−36 to −27 degrees Fahrenheit, −38 to −33°C]. The white regions appear to have temperatures on the order of 205 to 210 Kelvin [−90 to −81 degrees Fahrenheit, −68 to −63°C]. They cannot be carbon dioxide." At those temperatures, carbon dioxide would vaporize. "I believe we have now answered what has been the question of major controversy for about five years, and that is that the residual polar cap is made out of water ice. There's not a carbon dioxide reservoir in the polar regions."

This very important discovery was confirmed by measurements of atmospheric water vapor over the north polar cap in the experiments led by Dr. Barney Farmer. "From simply the water-vapor abundance we have to draw the conclusion that ice is predominantly water ice and not carbon dioxide ice," he said.

The total abundance of water vapor measured near the polar regions just before conjunction requires that atmospheric temperature close to the surface be higher than −99.4 degrees Fahrenheit (−73°C); lower temperatures would not permit so much water vapor to be present. These measurements, said Dr. Farmer, were incompatible with a polar cap of frozen carbon dioxide. The residual cap and the various patches of ice that surround it are mainly water ice. The thickness of this ice was estimated as somewhere between 3.3 feet (1 m) and 3,280 feet (1 km).

Viking also discovered that the atmosphere over the north pole is close to saturation with water vapor, so that snow falls there. But the total amount of water vapor observed over the polar regions was quite small. If it were all precipitated from the atmosphere onto the cap, it would only increase the thickness of the ice by about 0.004 inches (0.1 mm).

Viking mapped water vapor generally in the atmosphere. The biggest daily variations in precipitable water vapor seemed to occur at high elevations—from zero water at dawn to about 0.0004 inches (10 or 12 micrometers) at midday. In the high Syrtis Major region, water vapor increased very quickly after sunrise, possibly as surface frost flashed into vapor.

Middle latitudes showed random variations in the amount of water vapor. In some areas it decreased from dawn to sunset, while in others a slow increase was measured during the day. Still other areas showed no diurnal change. The apparently random nature of these effects may have been due to the presence of clouds. In general, high elevations on Mars had less water vapor than low elevations (figure 8.4), and water vapor was most abundant in a band around the planet at 70 to 80 degrees north latitude (figure 8.5).

The planet has two relatively low troughs that extend from the equator to the polar regions. As mentioned in the discussion about the landing site for Viking 2, one passes through Mare Acidalium, north of Chryse, and the other extends north of Utopia. Both areas show dark albedos, i.e., they appear as dark regions, on classical maps of Mars. These "drainage" regions showed the greatest changes in water vapor concentrations.

8.4 The amount of water vapor in the atmosphere of Mars at various longitudes around the planet is negatively correlated with the elevation of the surface. There is more water vapor above lower areas than above higher ones.

8.5 The team investigating atmospheric water vapor also discovered that the amount decreases toward the equator and toward the pole, with a maximum at about the edge of the polar cap.

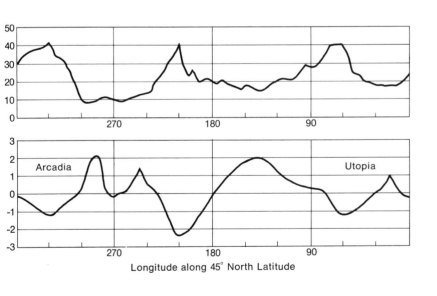

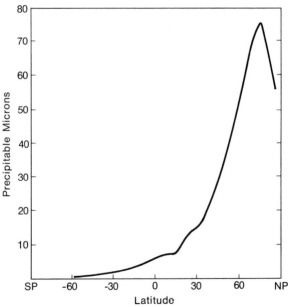

Generally the humidity is high in the Martian atmosphere; it contains about as much water vapor as is possible at the low temperatures. And Mars is a cloudy planet, many different types of clouds having been identified. Dr. William Baum of Lowell Observatory, who has studied Martian clouds for many years by telescopic observation from earth and was on the Viking orbiter imaging team, has classified the various types.

Hazes are dense blankets of condensation that are more prevalent in the northern hemisphere of Mars than in the south. They may be caused by condensation on small particles of dust, and they often show wave patterns of about 6 miles (10 km) wavelength. Such clouds are about equally visible in red and violet light, showing that the red light is not aiding in penetration to the surface beneath the haze; the particles making up the haze must be fairly large. The hazes extend high enough in the Martian atmosphere to obscure surface features, in contrast with low-lying fogs through which higher elevations penetrate.

Diffuse white clouds appear in the haze blankets from time to time. They appear to be brighter in violet light and are probably smaller particles than the haze particles. Lasting for hours, sometimes for days, they have been observed from earth for many years. They do not appear to move rapidly.

Large cloud systems related to surface features are associated with mountainous areas, particularly the Tharsis plateau, and with the volcanoes. These clouds appear during the morning as diffuse masses and then develop into more complex structures. These clouds have been observed for decades from earth and often appear as a prominent W-shape on the disc of Mars. They seem to be typical orographic clouds, which form on earth when masses of atmosphere are forced to rise as they move slowly from low-lying areas to mountainous areas and water condenses at the higher, colder altitudes. Some of those clouds rise almost to the summits of the volcanoes, about 12 miles (20 km) above the mean surface of Mars.

There are also convective clouds consisting of discrete cloudlets that move as groups, sometimes as fast as 200 feet per second (60 meters/second). These probably result from atmospheric gases being heated near the surface during the morning and rising higher. The expansive cooling condenses water to form the clouds. These clouds appear very frequently in equatorial latitudes about noon at a height of 3 miles (5 km) or so. These clouds do not cast strong shadows and do not change form rapidly.

Wave clouds on Mars were first observed in the Mariner 9 pictures. These clouds form when strong winds blow across obstacles such as ridges or crater walls. The clouds appear at the crests of the waves formed in the atmosphere by the obstacle to the wind motion. These wave clouds have been extremely valuable in our study of wind flows on Mars and have shown that the wind patterns are very similar to those in the same latitudes on earth. But the Martian winds move much faster at similar altitudes.

Bright patches in the bottoms of craters and channels were a surprising discovery by Viking. They showed that morning sun was turning ground frost into vapor that condensed into fog in the cold air. This was an exciting discovery because it showed that water was being interchanged between the Martian surface and the atmosphere each day.

Detached, thin layers of clouds are prominent in pictures of the limb of Mars, high above the haze layers (figure 8.6). They were discovered on the first picture of Mars taken by a spacecraft in 1964. The highest are commonly 25 miles (40 km) above the surface.

Some may be even higher on occasions. They are extremely thin and would appear to be transparent from the surface of the planet, looking up.

Polar hoods of clouds are very extensive in the northern hemisphere during winter. They reflect strongly in violet compared with red light and are probably small particles of carbon dioxide ice. There may even be snowstorms of carbon dioxide ice in polar regions during the winter there. Viking did not, however, find a polar hood over the southern cap as expected.

Local dust storms were recorded on some Viking pictures (figure 8.7). A global dust storm was not expected, nor seen, before solar conjunction. This type of storm does not usually develop until Mars approaches close to the sun at perihelion and its atmosphere receives the maximum amount of solar energy. Viking did observe some very large dust storms during its extended mission, storms that occurred sooner than expected. It is believed that the dust storms occur because dust-laden atmosphere absorbs more heat, and tidal winds are thereby increased to the point at which they can raise dust and further augment the heating of the atmosphere. The beginning might be when a local topographic wind adds to the tidal wind and starts the process operating. This appears to be the way in which dust storms begin in one area of Mars and spread rapidly around the planet.

Dr. Michael McElroy used the isotopic composition of today's Martian atmosphere to seek clues to the amount of water that might have been present on Mars in the past. Assuming a venting from the interior similar to that believed to have occurred on earth, Mars in its remote past would have had 54 atmospheres* of water, 5 atmospheres of carbon dioxide,

* One atmosphere equals 1,013 millibars, the average atmospheric pressure at sea level on earth.

and 0.1 atmosphere of nitrogen released from its interior. But from the Viking measurements of isotopes of atmospheric gases, Dr. McElroy calculated that there were initially only 2 atmospheres of water and 0.002 to 0.03 atmospheres of nitrogen. Mars may thus have had a very dense atmosphere in the past, somewhere between these upper and lower limits.

The relationship of argon 36 to argon 40 measured by the Viking experiments still remains a puzzle. Argon 40 is produced by the radioactive decay of potassium, while argon 36 comes from the original material forming the planet and provides an indication of the amount of outgassing from the interior that may have taken place. If the amount of volatiles on Mars was proportionate to that on earth, the amount of argon 36 relative to argon 40 suggests that Mars underwent less outgassing than earth, and the early atmosphere of Mars could have been about one-tenth that of the earth's pressure today. Alternatively, though less likely, perhaps Mars had no magnetic field and the noble gases could be stripped from its atmosphere by a strong solar wind.

The amount of oxygen also leads to some questions of origins. The time required for oxygen to escape from Mars into space is quite short today. But it could have been much longer in the past because nitrogen could have been more plentiful, and its presence would have slowed down the escape of oxygen through the topmost part of the atmosphere, known as the exosphere. As a result, a fairly dense atmosphere of oxygen might have accumulated on Mars in a period of a million years, as the gas was released from the rocks of the planet. Incoming solar ultraviolet radiation would have formed a layer of ozone, which in turn would have further restricted the escape of oxygen into space. Conditions would change only when nitrogen leaked into space. Thus there could have been at least two periods when the atmosphere remained unchanged for a relatively long

A.

8.6 Limb hazes are shown in *A,* a mosaic of pictures taken through a violet filter. The picture shows other types of clouds at left, and sinuous canyons filled with early morning mists. These canyons are shown in another view, from almost overhead, in *B. C* is an enlargement of the channels in the top right corner of *B*.

8.7 A local dust storm appears in this set of pictures. The left picture was taken in July 1976 and shows the area of the Solis Planum and the Labyrinthus Noctis, part of the Valles Marineris (*at top*). The right picture, from a slightly different angle, shows the same region on March 25, 1977, with a dust storm stretching about 375 miles (600 km) from west to east. The eastern (right) edge is sharply defined and is the front edge of the storm. Cellular structure in the dust cloud indicates a strong turbulence. There is some water ice among the dust, since the cloud shows up strongly in violet light.

C.

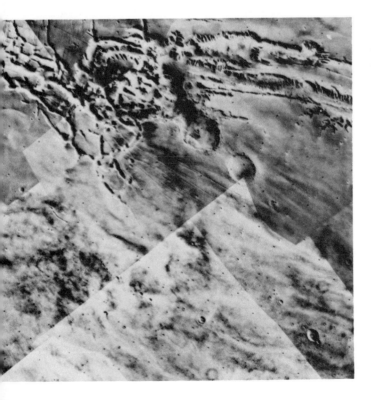

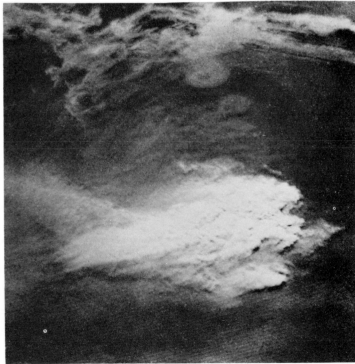

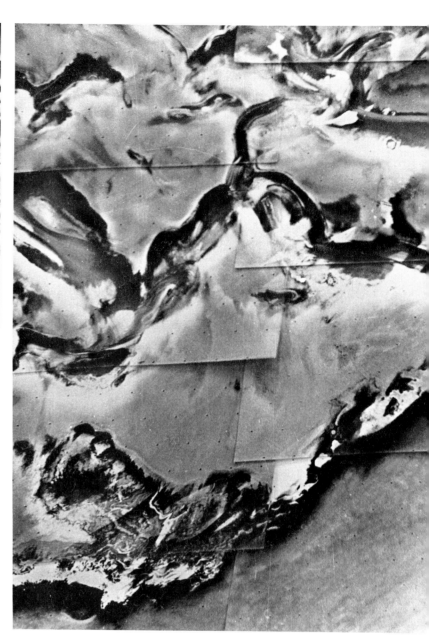

B.

A.

8.8 A type of feature never before observed on Mars was seen in this photomosaic, *A,* of the north polar ice cap, taken October 4, 1976 by orbiter 2. Translucent streaks of varied tones overlie both ice and defrosted layered material. They may be formed by redistribution of ice and soil particles by wind. The area shown extends about 225 miles (360 km) from 77 degrees north, 350 degrees west at the edge of the ice (bottom), to 83 degrees north, 330 degrees west (top). *B,* an enlargement of the edge of the cap at the bottom of *A,* showing the transition from ice to dune fields.

time. One of these was far in the past, when the atmosphere was rich in oxygen. Another is today when the atmosphere lacks oxygen. Volcanic eruptions might trigger a change from one state to the other. In the past, Mars might therefore have experienced a period (possibly several periods) when its atmosphere was much denser and richer in oxygen than it is today.

Interest in the Martian poles was stimulated by Mariner 9's relatively low-resolution pictures of the regions. A change in the orbit of Viking's orbiter 2 brought it over the north polar region at a low altitude, and it took high-resolution pictures of the area for the first time. These pictures (figure 8.8) showed three major types of terrain: layered deposits, extensive dune fields, and cratered plains.

Frost-free bands made the polar cap look somewhat like a pinwheel. These dark areas were clearly visible as a series of spiraling bands that were thought to represent places where ice had melted from terraced slopes (figure 8.9). The terraces had ice deposits on their flat surfaces, and their vertical surfaces were exposed. The terraces most likely represent layering in the subsurface rocks, owing to deposition of wind-carried dust, perhaps intermingled with ice, over millions of years. The layering may have resulted from climatic cycles caused by perturbations in Mars' orbit.

In some places the polar terraces crossed one another (figure 8.10). In geology this is known as an unconformable contact, of a buried landscape type. One speculation is that layers were deposited thickly enough to obscure all the existing topography of the polar region. The layers eroded, forming slopes with terraced edges as seen today. But in a second period of deposition a new set of layers covered the eroded first set. Later, erosion again dominated and carved more terraces, at a different angle from the earlier ones. There may, indeed, have been several periods of erosion and deposition.

Although most of the region had no impact craters and must be relatively young, a few ancient craters were embedded in the layers. Dr. James Cutts has suggested a scenario for the evolution of the polar regions. At stage one the polar regions looked like the rest of Mars, before any layering took place. At stage two the horizontally layered materials were deposited by the winds. In stage three, terraces formed on this layered material, probably by some erosional process. Stage four was deposition of a second batch of layers, and stage five was a further erosional attack that created the current set of terraces and sand dunes. Stage six is the present form.

Astronomical research suggests that Mars undergoes climatic cycles that are 50,000 years and 2 million years long. Each polar layer might be associated with a time scale of this kind. The Viking pictures showed that beyond these climatic cycles the climate has shifted at least twice in major ways, when the deposition changed to erosion and back again. At present scientists cannot say if the polar regions are depositing layers or eroding them.

What happened to the material that eroded to create the terraces? Viking's pictures showed details of the dark zone around the polar cap, which turned out to be the largest dune field in the solar system. (Earlier astronomers, looking through telescopes, had speculated that this zone might be water).

Some of the eroded material, said Dr. Cutts, appeared to have formed this 185-mile (300-km) wide girdle of sand dunes that surrounded the residual cap, stretching from 80 to 75 degrees north latitude. Dune fields issued from canyons that extended far into the polar ice cap (figure 8.11). In some places the dunes extended up over the layered deposits, thereby indicat-

Icefields Colored Red

141

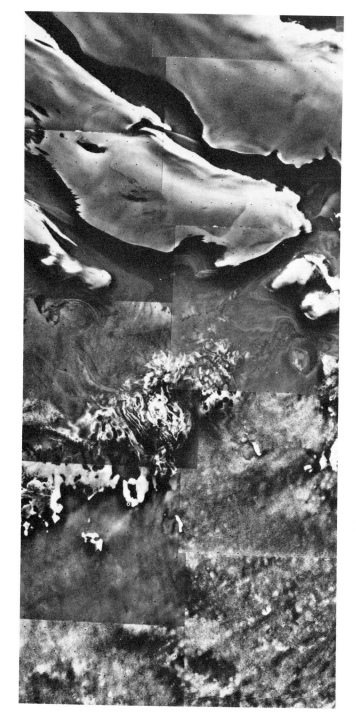

A.

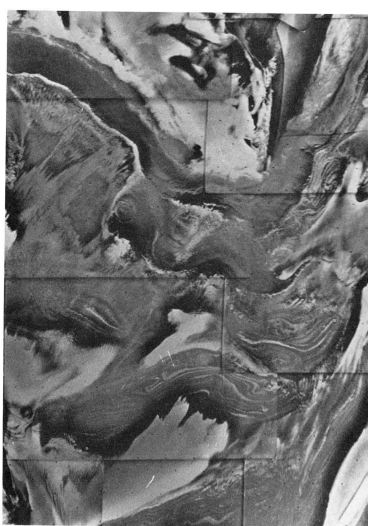

B.

8.9 The north polar cap, the layered deposits, and the surrounding dune fields are shown in *A,* taken by Viking 2 in November 1976. The terracing is evident as alternating light and dark bands. It is believed to record a series of cyclical climatic changes on Mars. *B,* a typical terraced area in the cap. This picture also shows some of the peculiar searchlight patterns which may result from the wind blowing ice and sand across the cap.

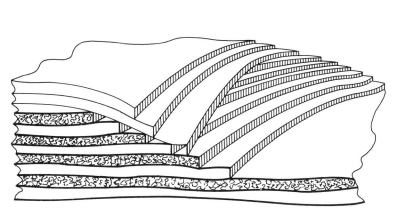

8.10 An explanation of the terraces in the polar cap proposed by Dr. James Cutts is that a first series of layered deposits was partially eroded and then covered by a new series of layered deposits, which was also later eroded. The discontinuities between the two sets of layered deposits might indicate a major climatic change on Mars.

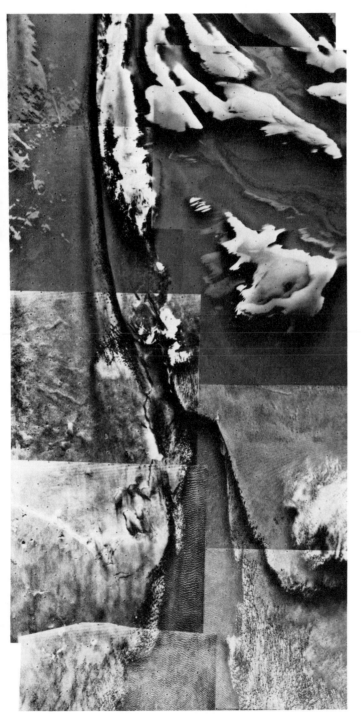

ing that the dunes are the most recent features. The layered deposits and dunes appear to have formed over the third type of polar terrain, the cratered plains.

Vast areas of the dunes were parallel ridges of very similar spacing; elsewhere dune fields were broken in placed by ridges that trended in a northerly direction. Geologists inferred that the Martian winds blew perpendicularly to the ridges, as on earth. This means that the winds moved in a circle around the pole. Some dunes formed separate linear structures, and individual dunes called borkans appeared to be migrating across the Martian surface away from the pole.

The other characteristic features of the north pole were its permanent ice and its seasonal ice. This pattern was thought to apply to the south pole too. The permanent ice was believed to consist entirely of water, whereas the seasonal ice contained both water and carbon dioxide. The permanent ice covered the major part of the layered deposits, and also occurred in patches round the cap, in dunes, and on the cratered plains. Viking pictures revealed a spiral pattern of ice concentrated on flat areas. The slopes were, however, generally free of permanent ice. The low reflectivity (albedo) implied that a lot of dust was mixed with the ice, and color pictures showed, in fact, that the ice fields were tinted red, probably with the ubiquitous red dust of Mars.

8.11 The edge of the polar cap, showing one of the canyons that penetrate into it. The floor of the canyon is covered with a regular dune field. Note that in this area of the cap and its surrounding dune fields there are hardly any impact craters, showing that the polar region is one of the youngest geological regions of the planet.

The polar ice seemed to be much thicker than was expected. The exact thickness is unknown—it may be anywhere from 10 meters to hundreds of meters, as estimated from the surface features it obscures. At the edges of the ice there were unusual features shaped almost like searchlight beams, which may be caused by wind action. The ice pack also had dark streaks in it.

Mysteries deepened elsewhere as observations poured in about the geology of the rest of the planet. The mystery of the "canals" of Mars changed to the mystery of the channels. Whereas 40 years ago or more, astronomers hotly debated the existence of a geometrical system of straight lines forming a network over the Martian surface, today it is clear that only a few surface features might have caused earlier astronomers to imagine this set of lines. Some features such as the Valles Marineris are visible by telescope from earth; in fact, this area corresponds to the classical canal Agathadaemon. Similarly the cliffs around the base of Olympus Mons are shown in some early maps as a system of straight lines forming part of the "canal" system. Shadings around Elysium were similarly interpreted, with the dark area extending from Cerberus forming part of a classical canal. The other faint dusky markings that appear on albedo maps showing the light and dark areas of Mars and, when viewed through half-closed eyes, take on the appearance of lineaments, do not match any surface features.

There were no canals in the classical sense, but Mariner 9 and Viking orbiters saw Mars as a planet of meandering channels that resembled old riverbeds, often with bright early-morning mists filling their floors (figure 8.12). Channels meandered over parts of the planet in great profusion, giving the appearance of major terrestrial river basins as seen from a jet at 35,000 feet or so. Many channels looked like the great erosional chasms of the earth in arid and arctic lands. Yet if water eroded these channels,

where did it originate, and where is it today? These questions nagged at the Viking scientists and others who studied the pictures from Mars.

Almost every picture of the equatorial regions of Mars had channels (figure 8.13). Even the two landing sites showed evidence of fluvial action. There were prominent dendritic (branching) channels west of the Chryse landing site, forming spectacular channels on a large plain (figure 8.14). An ancient flood had swept across an area 125 miles (200 km) wide. Wherever there were barriers to the flood it funneled through gaps, gouging deeply as it went. In several places the flood appeared to have broken through one side of a crater rim, filled the crater, and then burst out on the far side.

The flood apparently began in large areas of collapsed terrain southwest of Chryse in an area known as the Lunae Planum, from which channels of enormous dimensions originate. It turned right through old cratered terrain, and it flowed through the mountain walls of Chryse and onto the plain. In Chryse the dimensions of the flood were much greater than anything known on earth. In Lunae Planum, craters diverted the flow and gave rise to flow lines. In Chryse there are similar flow lines and clear evidence of streamlining.

The collapsed terrain (figure 8.15), often called chaotic terrain by the investigators on the project, exhibits a rough floor of large, jumbled, angular blocks, separated from nearby plateaus by escarpments or cliffs up to 2 miles (3 km) high. Subsiding and slumping seem to have played a major part in creating this terrain, and melting of subsurface ice may have been the cause.

Viking also showed many channels originating in the edges of the Hellas basin and running down into it. In one channel the details of the floor suggested that it might be a flood plain cut by a much smaller stream, the meandering of the stream having resulted in the wide channel. There are short furrows on sloping ground everywhere throughout the old cratered terrain, suggesting drainage patterns from rainfall. Another type of channel originated in the area of chaotic terrain. These channels were much wider. They did not have any area of water collection, but seemed to result from a single source of water in the ground. Several appeared to flow from the base of cliffs.

A most common type of channel was one that mysteriously started nowhere and ended nowhere. These are somewhat like rivers in the western deserts of the United States and probably arose from similar processes—local precipitation or springs and evaporation into a desert at the lower end.

Dr. Carr and his team have tried to estimate the age of channels by counting the numbers of craters on them. Such crater counts are most easily made on the tributary systems, where they produce about the same percentages of craters of different sizes as counts of craters on the big volcanoes. The freshest tributary systems have ten times as many craters as the volcanoes. The channels therefore are much older than the volcanoes, and volcanic outgassing is accordingly unlikely to account for the floods. The tributary systems, however, have fewer craters than do lunar maria (the floors of the biggest impact basins on the moon). This may mean that the tributary systems are less than 3 billion years old if the periods of cratering and rates of cratering were the same on Mars as on the moon.

Where did the water come from? While some small channel systems might be explained by heavy local rainfall, nothing indicates that it rained generally on Mars. There is abundant evidence for ground ice on Mars, so the water might have originated from its melting. A climate change could do this, or geother-

mal activity (the upwelling of hot magma from the interior of Mars toward the surface) might have caused it by melting ice mixed with the surface rocks.

Many scientists accepted the channels as evidence for extensive fluvial periods (periods of water flow) on Mars, but others warned that a theory of water flow might be leading down a blind alley. Dr. Cutts pointed out that the amount of water stored in crustal rocks was 1,000 times too small to carve channels of such size. The volume of soil excavated from the channels was greater than the volume of collapsed terrain from which the water was supposed to have originated. Also, the outflow channels from the supposed source were not diverted by crater rims but bisected them with deep gorges, sometimes many miles from the main channel. It is difficult to explain how a flow of water could breach the crater rims to start cutting the deep channels through them.

There is no evidence that Mars ever had standing bodies of water, ancient oceans or seas, because we see no shorelines. But there are sediments and lava flows that might have covered such shorelines. A Martian ocean might even have frozen over and not produced the terrestrial-type shorelines for which we look in vain on the Viking pictures. What might be an alternative to water?

All the channels might have originated as endogenic valleys created by forces from within the Martian crust, by tensional forces, subsiding of the soil, and volcanism. Winds might have later been focused by these valleys to cause the scouring and streamlining in the valley floors. Winds could do this over billions of years. Moreover, tensions in the crust could create a forked fracture system which, after further subsidence and possible wind erosion, would result in the branching patterns seen on the planet. The channel system might also result from a combination of wind and water action.

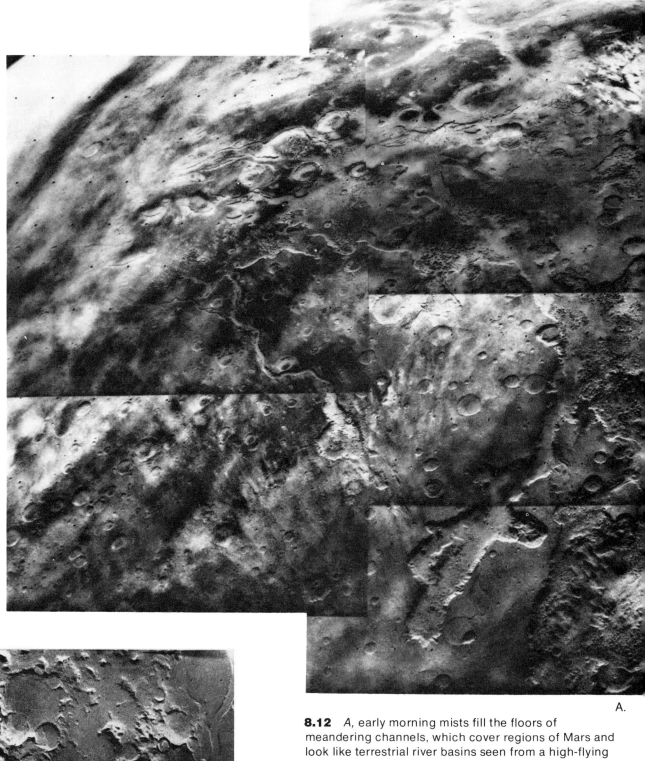

A.

8.12 *A,* early morning mists fill the floors of meandering channels, which cover regions of Mars and look like terrestrial river basins seen from a high-flying jet plane.
B, close-up of one of these sinuous channels in a vertical view. Note the mountain mass with the erosional feature surrounding its base.

8.13 On close inspection, the heavily cratered terrain of Mars has numerous channels that appear to have been formed by running water, possibly from local rainfall. Some channels, e.g., those at the right-hand side of *A,* appear to start and end nowhere; others, such as the one down the center of the picture, continue for great distances and have many tributaries. *B,* the slopes of the crater walls are covered with drainage channels that look more like water channels than lava channels.

B.

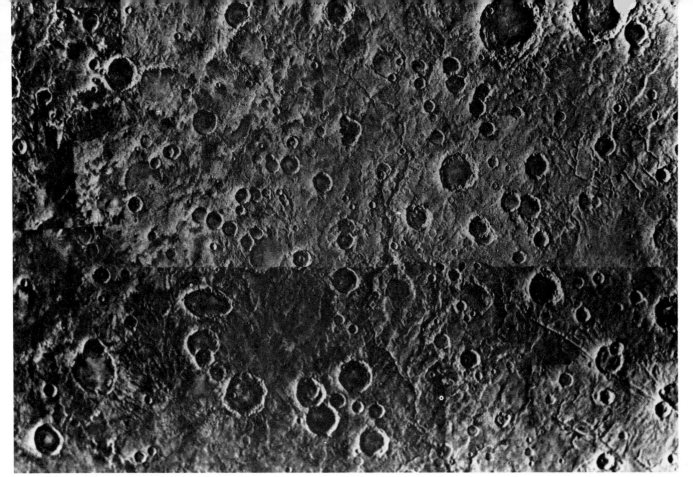

8.13 A.

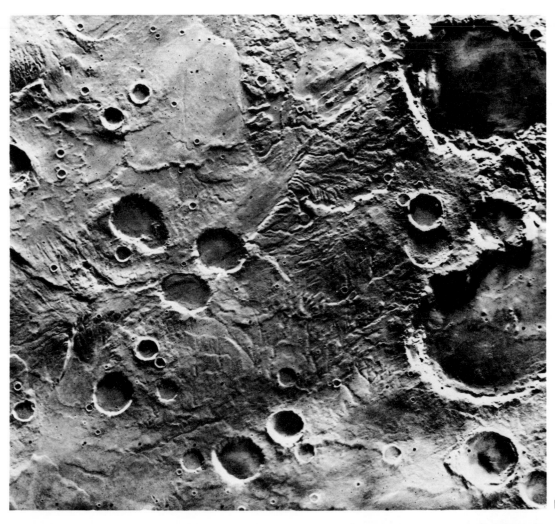

B.

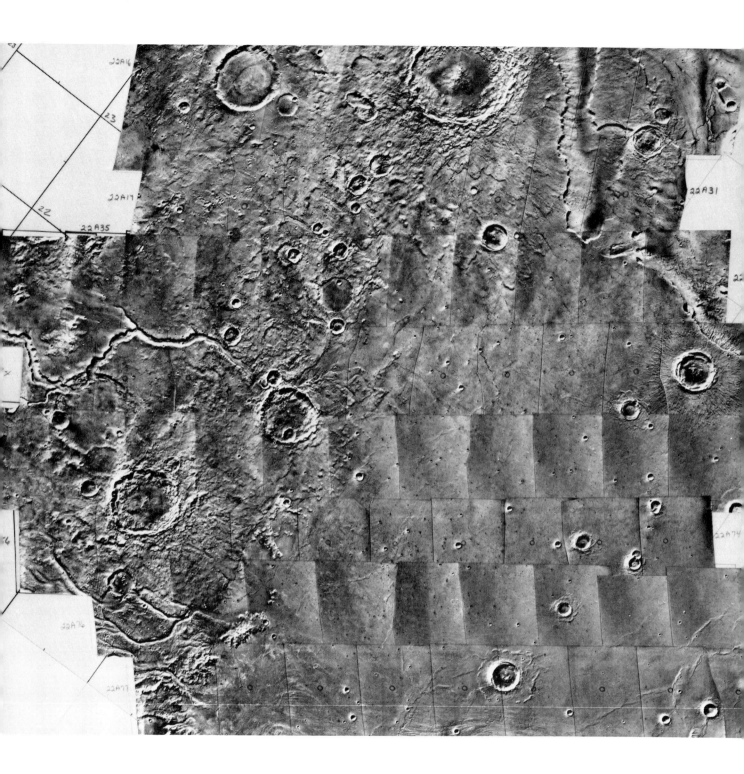

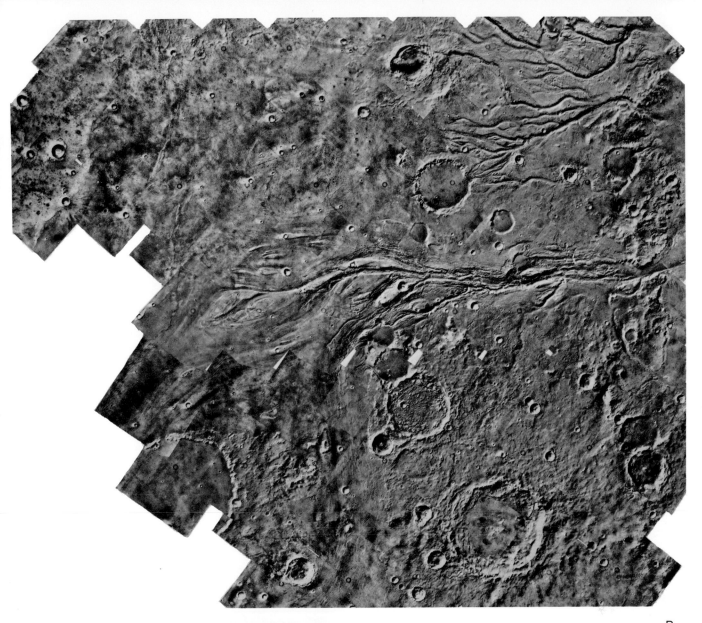

B.

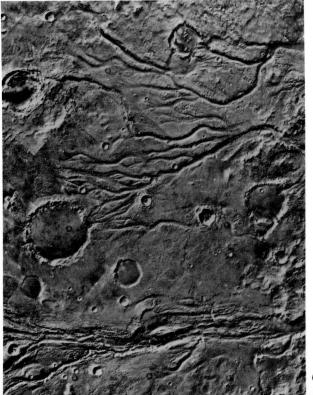

8.14 *A,* a mosaic of the northwestern boundary of the Chryse basin shows the channels entering it from Lunae Planum, and wide erosional features with high cliffs at the upper right. *B* continues the details of the boundary southward along the west of the the planitia, showing mainly the Lunae Planum. *C* a high-resolution mosaic of the top right part of *B.* It is centered at 17 degrees north latitude, 55 degrees west longitude, to the west of the landing site for Viking 1. The terrain slopes in this picture from west to east, left to right, with a drop of about 1.86 miles (3 km). The channels suggest a massive flood from Lunae Planum into the Chryse basin, possibly originating from an area of chaotic terrain at the southern part of Lunae Planum.

C.

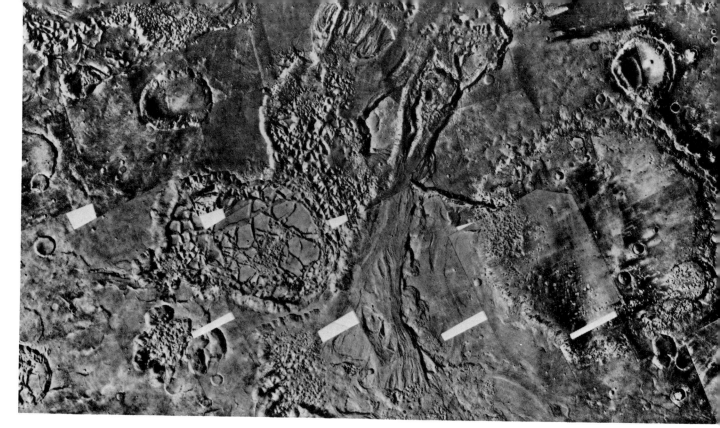

8.15 Photomosaic taken by Viking 1 near the eastern end of Valles Marineris shows several large collapse features associated with outwash channels and chaotic terrain of jumbled blocks. Scientists believe that a possible explanation for many collapse features is subterranean melting of ice and subsequent draining away of the water, in a northerly direction toward the top of this picture.

One possibility is that heat from deep inside the planet was conducted toward the surface and melted ice in the crust into underground pools that were dammed for a while and then broke out as massive floods. The hot water would then melt more water from the mixture of water and rocks and scour away large amounts of material from the flood channels as the water flowed to lower levels. This would have been analogous to pouring rivulets of hot water over a driveway covered with hard-packed snow and ice.

Another theory for the origin of the Martian channels was suggested by Ernest Schonfield, a lunar scientist of NASA's Johnson Space Center. He believed that the channels were easier to explain by lava action than by water action. He proposed that thin, low-viscosity, basaltic liquid melted beneath the planet's surface and flowed freely to erode the surface materials.

The amount of water on Mars is undoubtedly modest by terrestrial standards. If Mars outgassed only one-hundredth of that of earth, there would have been about 42 feet (13 m) of water over the whole of the planet. If Mars had the same composition as car-bonaceous chondrites (thought to represent the primitive materials from which the rocky planets were formed), it would be expected to have more water than earth. Then there should have been large amounts of water on Mars. But where?

The polar caps, now understood to be mainly water, do not store enough. They would only supply about 6 feet (2 m) of water over the whole of the planet if all their ice was melted. Only 3 feet (1 m) of water could have escaped from the exosphere into space during the planets's evolution. Ice might, indeed, have been the agent for carving the channels. These could not be ice canyons, however, because ice does not have long-term strength; it deforms and loses its shape. To store large volumes of ice and maintain topographic features for many millions of years would require

Icefields Colored Red

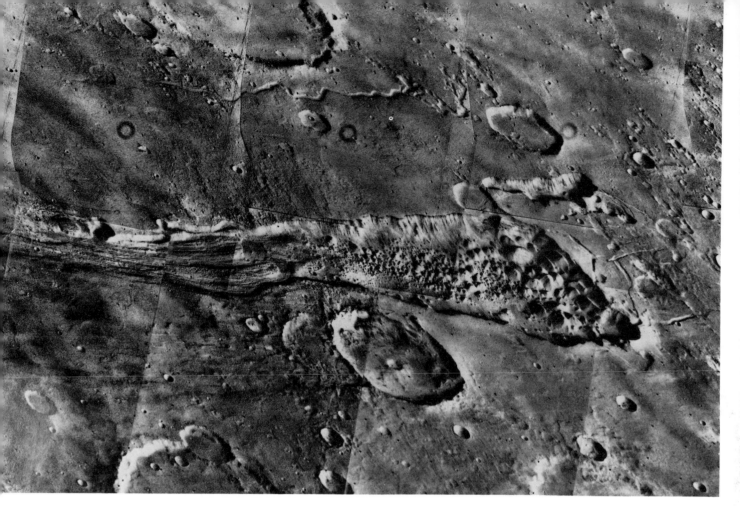

8.16 The foreground of this picture (taken by Viking orbiter 1 on July 3) shows a valley that was probably caused by downfaulting of the Martian crust, possibly by geothermal melting of a rock–ice mixture. The resulting water appears to have flowed out from this mini-basin to the left. The area covered by the picture is about 180 by 180 miles (300 by 300 km).

grains of rock in the ice for strength. These grains would have to be touching each other to provide the necessary strength. Hence there could only be about 30 percent of ice present in the regolith, the loose surface rock, if its strength was to be great enough to preserve the surface features seen on Mars today.

Self-destruction of such ice-saturated strata could account for the erosional features seen on Mars (figure 8.16). Plateaus might be ice-saturated layered terrains that melt at their edges and leave behind rocks to form the surrounding rough areas of chaotic terrain. The action might have been analogous to desiccating a terrain of its cement (the ice) and allowing the remaining rocky material to subside. If the crust of Mars were saturated with ice, that would also explain the peculiar sheets of ejecta around many craters, which look like slurries of rock and water (figure 8.17).

Ice could have been important in three ways to the molding of the Martian surface. The melting of a permafrost of buried ice would let the overlying region collapse and would release water to the surface, thus allowing the surface to be modified by flowing water. Glaciers might have flowed on the surface in the classical way that they have on earth. But there is no conclusive evidence of glacial processes on Mars; no moraines have been identified in the pictures from orbit. These piles of rocks carried by glaciers and deposited as the ice melts are characteristic of terrestrial glaciers. Thirdly, the ice may be a rock type. Martian crustal rocks might be a combination of sedi-

8.17 Base surge craters, whose material appears to have splashed out like mud across the surrounding surface, are common on the northern plains and elsewhere on Mars. The peculiar form of the ejecta blankets shown on these two pictures (taken in Cydonia) may have resulted because the meteors fell into wet sediments of a shallow body of water, or into a mixture of rock and ice.

ment and ice. When the ice melted, the sediment would flow as mud does on earth.

Dr. Harold Masursky, of the orbiter imaging team, concluded that the dendritic channels were caused by rainfall and the big channels by melting ice. He pointed out that some dendritic channels on the flanks of the big volcanic plateau of Alba Patera have jogs caused by faults offsetting them, just as the San Andreas fault displaced stream channels in the San Joaquin Valley into two abrupt bends where they cross the fault line. Dr. Masursky counted craters on several Martian landforms. Interestingly he found that the counts were almost the same for a channel near the first Viking lander site and for the Alba Patera. Using the crater counts, he established some relative ages of Martian landforms that may be related to lunar ages (see figure 8.18).

Crater counts also showed that there has been volcanic activity on Mars throughout the history of the planet, and tectonic activity too. The channels appear to be of all ages. Thus, Masursky claims, ice melting, rainfall, and channel formation happened many times over Martian history.

Other Viking scientists say, however, that although the evidence for lava flows and impact craters is very clear, that for flowing water is not so clear nor is it generally accepted. An outstanding conflict before Viking was the contradiction between the fluvial channels—first seen in the Mariner 9 pictures—and the present dry state of Mars. Viking has not resolved this conflict.

Icefields Colored Red

Relative Ages of Martian Features Established From Crater Counts

Feature	Billion Years
Tharsis and volcanoes	0–0.5
Young plains	1.0
Tharsis fractured plains	1.1
Oxia Pallus	2.5
Large upland craters modified	2.5–3.5
Old plains	3.5
Heavily cratered terrain	3.9

8.18 Table based on crater counts by members of the orbiter imaging team attempts to place relative ages on various Martian features. As with trying to establish ages for features on the moon, such estimates can only be made firm by obtaining samples of the Martian rocks and dating them with the various techniques now available. The ages in this table must be taken as being very tentative.

All crater counts seem to indicate that the fluvial period when the channels were cut preceded the formation of the Tharsis volcanic ridge. The flanks of this ridge have no channels, which supports the finding of the crater counts. There are channels, however, on the volcanic region of Alba, and crater counts show that this region is more ancient than Tharsis.

There are three types of volcanic features on Mars: Olympus Mons and the other young volcanoes of the Tharsis region, older volcanoes such as Alba Patera, and the volcanic plains. The Viking pictures show lava flows superimposed over the lines of cliffs forming boundary scarps around Olympus Mons (figure 8.19). The flows poured over the cliffs and extended into the northeast and southeast plains. The relationship between the plains-forming, scarp-forming, and shield-forming processes is complex. The shield consists of rather gently sloping lava flows that form the bulk of the volcanic mountain, like the flows of Mauna Loa in Hawaii.

Olympus Mons is the largest shield volcano known. The whole structure is about 370 miles (600 km) across at its base, and rises about 17 miles (27.4 km) above the surrounding plains. The summit caldera (figure 8.20) is a complex feature recording a series of eruptions in varied levels of frozen lava lakes. The crater walls cast shadows that have been used to measure their depth. These walls are 1.5 to 1.7 miles (2.4 to 2.8 km) high and have slopes of about 32 degrees. The overall caldera complex is about 45 miles (70 km) across; the youngest of the caldera pits is about 19 miles (30 km) across. Scientists have speculated that the Martian volcanoes were able to grow to such great heights because the crust of Mars had not moved, as the earth's crust has. When a volcanic vent developed, it continued to amass lava flows for millions of years, an effect like piling all the islands of the Hawaiian chain on top of one another.

Alba Patera (see figure 5.10) is unique to Mars; there are no volcanoes like it in size on earth, the moon, or Mercury. It consists of a central caldera complex with surrounding scarps. Viking revealed extremely rugged terrain and showed that the feature is not nearly as smooth as it appeared on the Mariner 9 pictures. Individual lava flows extend 500 miles (800 km) beyond the peripheral circular fractures and over gentle slopes.

Volcanic activity seems to have alternated with the fracture-forming process that split the Martian crust. Eruptions appear to have been sporadic. Different flows exhibit different styles of volcanism, e.g., ridge structures with lava tubes running along the tops of the ridges, and basalt flows to as much as 250 miles (400 km), which is extremely long by terrestrial standards. Other flows were rapid-flowing flood sheets. A third kind of flow produced individual channels that indicate sporadic eruptions. All the flows seem to be of fine-grained, heavy basaltic lavas, rather than coarse-grained and lighter granitic lavas.

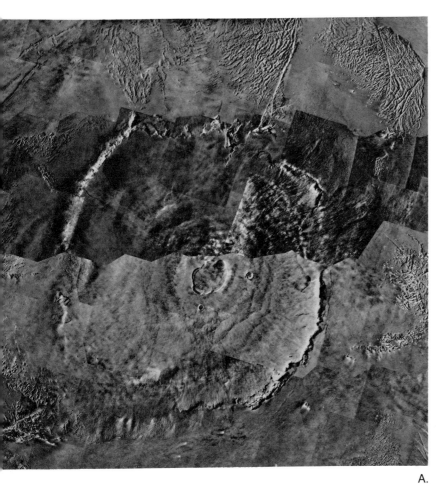

A.

B.

8.19 Olympus Mons is the largest known volcano in the solar system. *A,* general mosaic of the huge mountain and its surrounds, relatively clear of clouds. *B,* lava flows on the flanks of this shield volcano have sometimes almost obliterated traces of the base cliffs, themselves unexplained. The transition from the lava flows across the cliffs to the surrounding smooth plains is quite remarkable. This view is to the northeast of the shield. *C,* the summit caldera is revealed in this high-resolution mosaic. *D,* faults and peculiar mountainous terrain stretch north across the wide plains. The cliffs at the base of Olympus Mons are at the bottom edge of this mosaic.

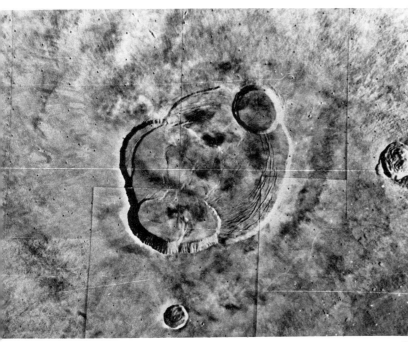

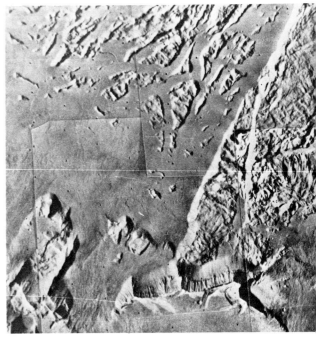

C.

D.

8.20 This close-up of part of the summit caldera of Olympus Mons was obtained by Viking 1 in June 1977. (North is at bottom, for ease in viewing.) The picture has five times better resolution than earlier pictures of this mountaintop, and shows features as small as 60 feet (18 m) across. It shows that the summit caldera is a complex feature in which a series of eruptions is recorded in lava lakes. Compare with figure 8-19C, which was the best previous picture of this caldera.

The old volcanoes on Mars, as evidenced by much erosion and higher crater counts (figure 8.21), are smaller than the young volcanoes. This suggests that the Martian crust was thinner in the past and could not support big volcanic piles as it does today. The remarkable fluvial activity etched into the Martian surface might have been associated in some way with a thinner crust that led to more geothermal heating of the rock–ice crustal rocks of Mars. Some estimates are that the Martian crust today must be about 100 miles (160 km) thick, compared with the 20-mile (32-km) thick crust of earth. But without good seismic data from Mars, the crustal thickness is very speculative.

The entire Tharsis region and its big volcanoes are fascinating. On the northeast of Arsia Mons (figure 8.22), the orbiter imaging team discovered a very large flow of debris, as though the sides of the volcano had slumped and slid out over the adjacent plains for a distance of several hundred miles. Similar features are recognized on Pavonis Mons and Olympus Mons, and they seem to be a characteristic of these big shield volcanoes. On a much smaller scale, the Hawaiian volcano Moana Loa has similar flows of debris.

Dr. Michael Carr described another important discovery connected with the south side of Arsia Mons, where plains extend for about 600 miles (1,000 km). These plains appear to be covered by lava flows from the main volcano. As distance increases from the main caldera, the number of impact craters superimposed on the lava flows becomes larger. This probably indicates that the more distant flows are older, said Dr. Carr. The materials immediately around Arsia Mons are likely to be much younger than those farther away, and the volcano has most likely been growing for a very long time, perhaps as long as two billion years.

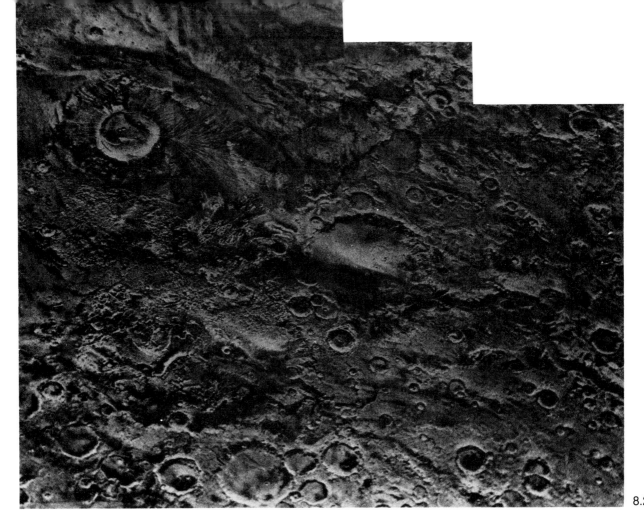

8.21 A.

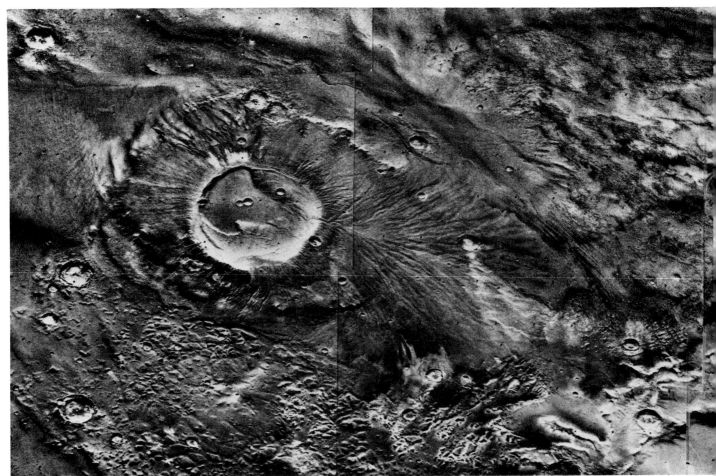

B.

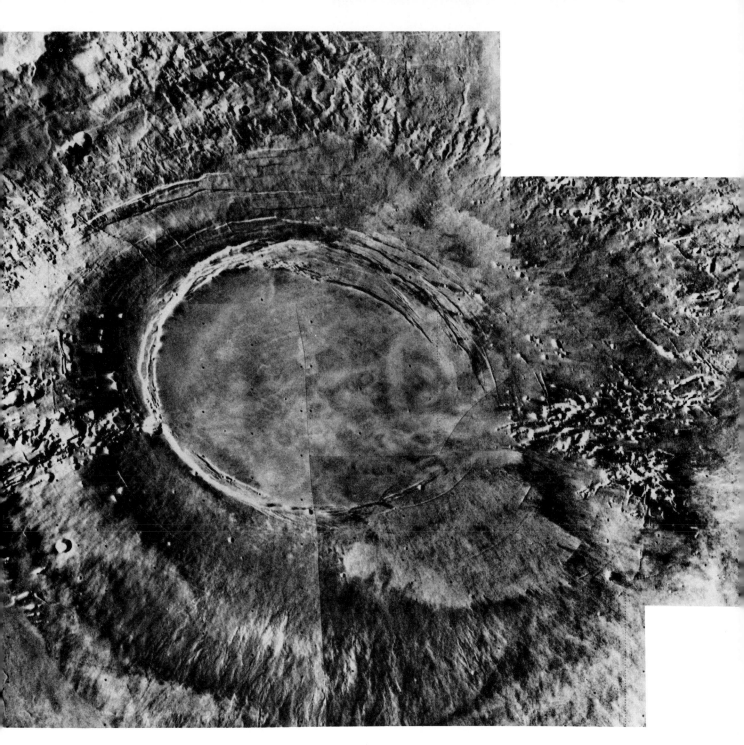

8.21 Apollinaris Patera is located at about 8 degrees south latitude, 186 degrees west longitude, far from the major volcanic regions of Mars. This ancient volcano has a central caldera that is 62 miles (100 km) in diameter, surrounded by shallow sloping flanks that terminate abruptly in a cliff. The number of craters on the volcano suggest that it is old. *A,* a general mosaic of the area. *B,* a higher-resolution picture of the volcano itself.

8.22 Arsia Mons is one of the three large Tharsis volcanoes. It was once called "South Spot" and stands 12 miles (19 km) high. Its central caldera is about 75 miles (120 kms) across. Outside of the caldera, numerous lava flows show fine linear features. Vast amounts of lava flooded surrounding plains.

Mars appears to have had volcanic activity on its surface for a major part of its history, and that activity might be continuing today. This can only be determined by measuring seismic activity on Mars because it is unlikely (based on terrestrial events) that a hot lava flow could be measured from orbit during the two-year extended mission of Viking. On sol 60, just before solar conjunction, lander 2 did record an event that could be interpreted as seismic. It must have been quite a large event if the sequence of waves is interpreted as resembling those from terrestrial earthquakes. Another similar event was detected a short while later. Mars seemed to be active seismically today.

The third type of volcanic landform is the volcanic plain. Many plains on Mars were formed in this way, similar to the lava flows on the moon's Imbrium basin. Relatively fast-flowing lava of flood eruptions produced sheets that encroached on craters and highlands. The flows draped around features, filled fractures, and covered ridges. Floors of some craters also welled up with lava and later developed polygonal cracks.

Another fascinating area of Mars explored by Viking from orbit was the great valley. This consists of a series of troughs and canyons along a broad ridge 2,500 miles (4,300 km) long, extending from Tharsis to Chryse. There are three distinct types of valleys in the system; a network of graben (depressed segments of the Martian crust bounded on either side by faults) named Labyrinthus Noctis (figure 8.23), a system of broad troughs named Valles Marineris (figure 8.24), and transitional troughs leading to the plain of Chryse.

The Viking photographs confirm that Valles Marineris is really a huge graben and that it contains complex sedimentary deposits. The canyon walls are deeply cracked and often truncated into steep walls by other faults near the bases of the cliffs. The general sequence of their formation seems to be that the graben formed and was then eroded and truncated at the base of its cliffs by later linear faulting. Still later, the canyon floor warped and became broken into parts (figure 8.24C). It also appears that part of the material moved downward by faulting.

The walls of the big valley display complex light and dark horizontal layers of materials, which suggest a complex earthlike geology on Mars. Alternate layers of light and dark bedded material of almost equal thickness can be seen in many places on the big canyon's walls. At one place, 22 layers can be counted in Ganges Chasma. These regular layers may indicate Martian climatic cycles, or wind-blown deposits stretching back for eons. The uniformity of the layers strongly suggests climatic activity.

The canyons also contain dune fields on their floors (figure 8.25), and they exhibit many landslides. Thin debris sheets often spread from the base of the landslides (figure 8.25A). Although there has been volcanism on Mars during the last 100 million years, there is no evidence of basalt flows within the canyons.

Viking also produced many pictures of the floor and environs of the Hellas basin (figure 8.26) that were pieced together for a first view of this mysterious area, one of the deepest depressions on Mars. Hellas is thought to have been formed by the impact of a mountain-sized body on Mars in the early days of the solar system. Its floor is some 3.7 miles (6 km) below the mean level of the Martian surface, and some parts may be deeper. During Mariner 9's survey of the planet, this floor was hidden by dust or haze and appeared featureless.

Viking orbiter 1 was able to take pictures for the photomosaic when the Hellas basin floor was clearly visi-

A.

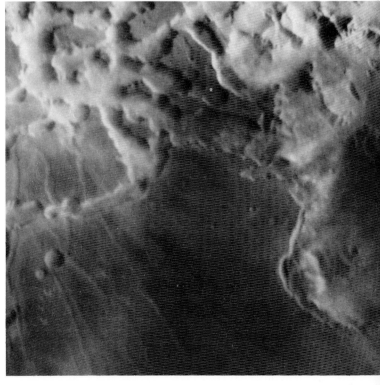

B.

ble. Light markings may be frost, but the pictures revealed flow patterns and craters on the floor of the basin, together with some very large linear structures. Lava flows from old volcanoes on the rim of the depression stretch down the walls into the basin. An important feature discovered in the pictures was fan-like forms that emanate from the base of features of higher elevation. Wherever there is a mound in Hellas, there is a fan of debris at the bottom of the slope. This is thought to be similar to debris flows found at the perimeters of glacial regions on earth. It reinforced the belief that there is much ground ice on Mars.

Another large impact basin was photographed many times by the orbiters (figure 8.27). The Argyre basin is not so deep as Hellas but its floor too is often obscured by mists and covered by frosts.

8.23 The first part of the canyons of Mars is Labyrinthus Noctis. *A,* a general view of the area, fairly free of mist within the canyons; however, there are many small clouds over the surrounding tablelands. *B,* early morning in the area shows bright clouds, most probably of water ice, in and around all the canyons.

One possible cause for these clouds is that water which condensed during the previous afternoon on shaded eastern-facing slopes of the canyon floor may be vaporized as the early morning sunlight falls on these same slopes.

A.

8.24 The mighty Valles Marineris is an impressive feature of the equatorial regions of Mars. *A,* more than 100 individual pictures form this photomosaic, which is centered at 5 degrees south latitude, 85 degrees west longitude. At left is part of Labyrinthus Noctis, from which two major canyons, Tithonium Chasma (*top*) and Ius Chasma, stretch to the east. See also figure 8.7, which shows an oblique picture of this same area with mists filling the canyons. *B,* a mosaic of the west end of Valles Marineris provides an oblique view across Tithonium Chasma (*top*) and Ius Chasma. The erosion of the canyons formed great slumps on the walls and cut side valleys seen clearly on this picture. *C,* details of the canyon floor, many layers and great fans of debris. *D,* further details of the movement of materials on the canyon floor. *E,* a major landslide from the wall of the canyon. *F,* unusual erosion of a crater on the rim of the canyon. This crater had a base-surge ejecta blanket.

Icefields Colored Red

8.24 B.

Icefields Colored Red

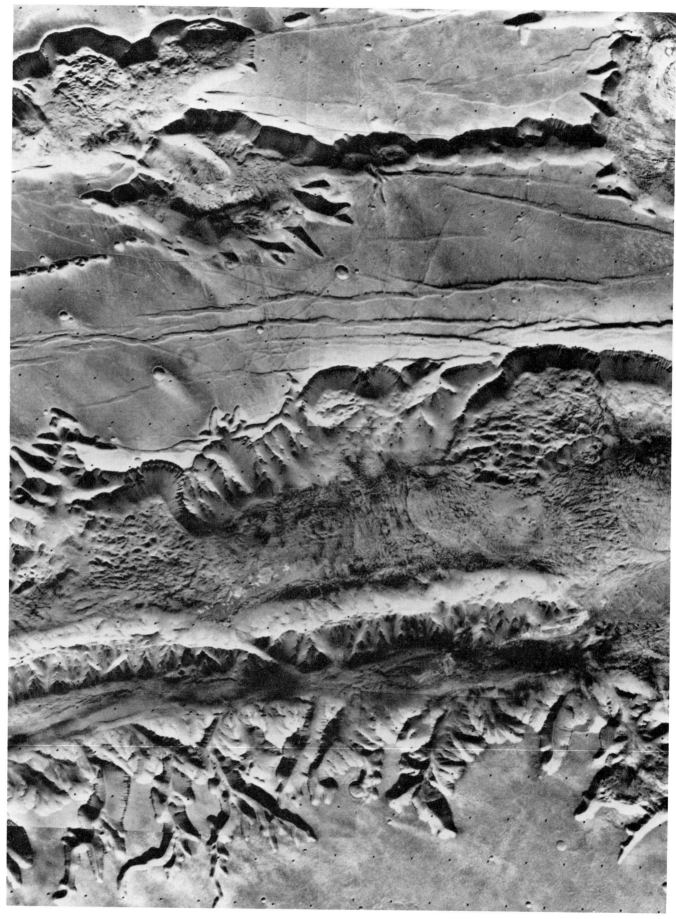

8.24 C.

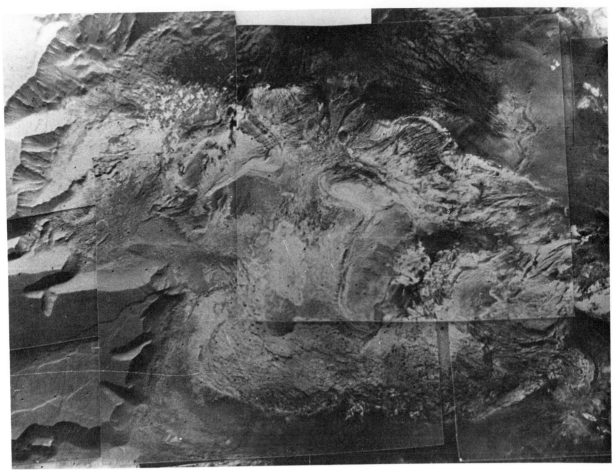

8.24 D.

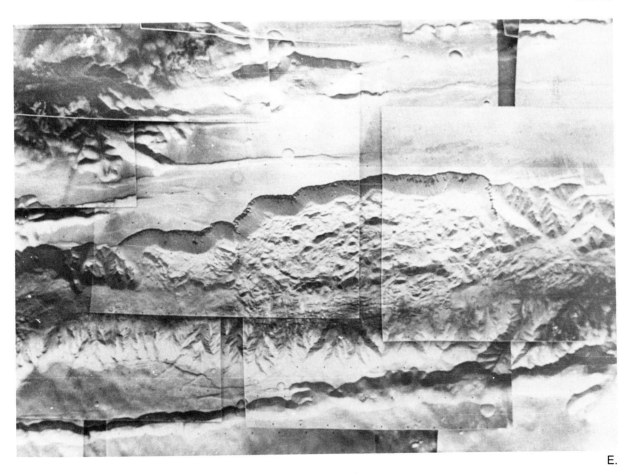

E.

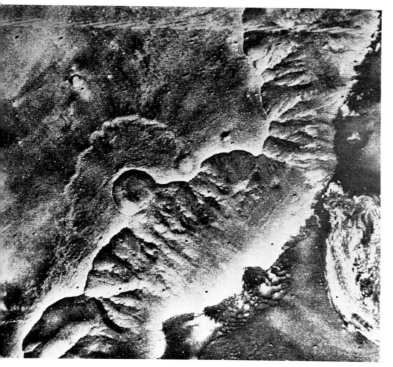

8.24 F.

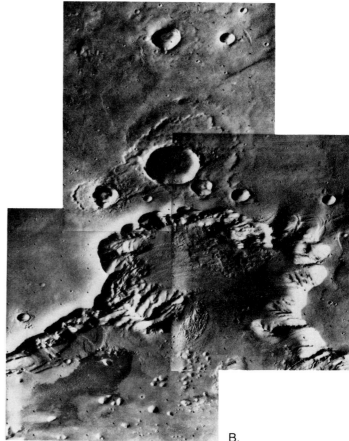

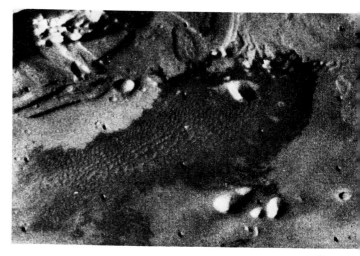

B.

C.

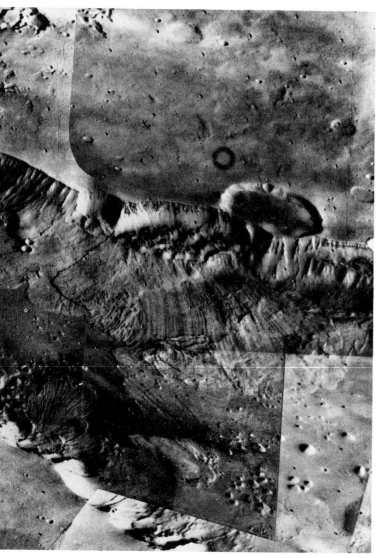

A.

8.25 As the canyons change into transitional troughs leading to Chryse, many interesting features appear on their walls and floors. *A,* in this view looking south across Valles Marineris, aprons of debris on the canyon floor indicate how the canyon enlarges itself from the original fault. The walls appear to collapse at intervals to form huge landslides that flow down and across the canyon floor. Striations show the direction of flow. One landslide is seen to have ridden over an earlier landslide on the far side of the canyon. *B,* in this view of the north wall of Ganges Chasma, erosional fluting can be clearly seen together with landslides. While the surrounding plateaus are dotted with impact craters, the canyon floors and landslides are remarkably free of them. The dark area at the bottom left of the picture, shown in detail in *C,* is a dune field.

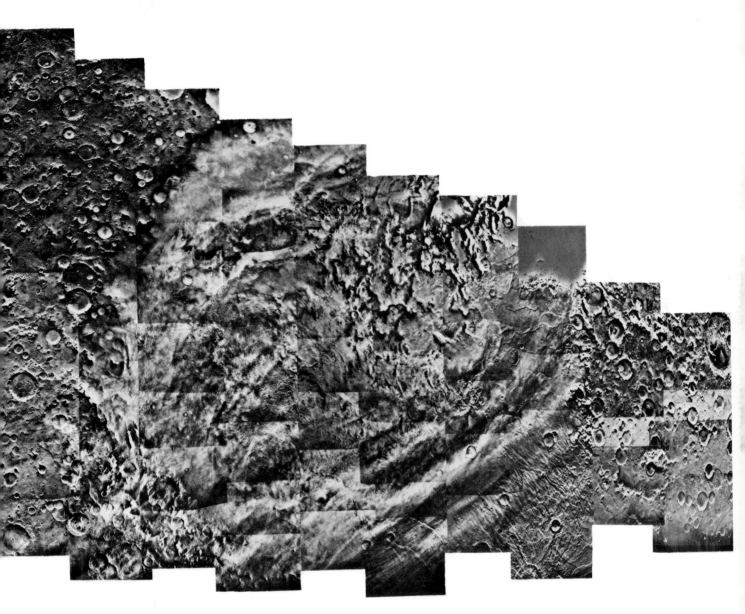

8.26 This mosaic of the Hellas region of Mars was taken by Orbiter 1 during September 1976 at a range of about 5,900 miles (9,500 km). This was the first time that the floor of this great impact basin had been seen. The basin is 1,100 miles (1,800 km) across and is 3.7 miles below the mean surface level of Mars. Frost on the floor results in striking albedo patterns. Many volcanic flows are visible, especially on the southern rim where there is an ancient volcanic crater on the bottom edge of the mosaic. Parts of the basin boundaries are obscured by clouds.

Icefields Colored Red   165

Mars, it seems, is a frozen world. In the words of Dr. Barney Farmer: "Mars, . . . in terms of its atmospheric water vapor, behaves just as one would expect a shell of ice to behave. The shell of ice is covered predominantly in the equatorial regions by some rock, and the cyclic annual variations [of water vapor] that we see are simply the [result of] changes of temperature of the covering rock with the changes of solar latitude. Putting this together with the north polar ice, we see that Mars can perhaps be described quite nicely as an iceberg. The north cap is the tip of the iceberg, and it's floating in this sea of rocks."

And over all the planet is the ubiquitous limonite dust, making it a planet of dusty red ice. That may be where the water is hidden, bound up as ice in this rock–ice mixture of the Martian crust. In the North Slope of Alaska, the permafrost extends to a depth of about 2,000 feet (600 m). In the crust of Mars, it might extend for many miles.

One thing is certain—our view of Mars following Viking is entirely different. We have discovered from the Viking expedition that Mars is unique unto itself, and quite unlike the earth, the moon, Mercury, or Venus. Interpreting its geology will be a long, hard, but fascinating task in comparative planetology.

A.

8.27 Another major impact basin in the southern hemisphere of Mars is the Argyre basin shown in these pictures. *A,* a wide sweep of the globe of Mars looking toward the south pole. There are high-altitude hazes over the limb of the planet, and streaks of clouds in the foreground. Many of the craters are filled with mist or frost. *B,* details in the wall of the impact basin, which is here filled with frost and appears very bright.

Icefields Colored Red

B.

Icefields Colored Red

Chapter 9

Epilogue

Although from the Viking expedition we cannot state categorically that there is no life on Mars, nor that life does exist on the planet, the mission clarified many other questions about our neighbor world. In the coming years the immense quantities of data returned by the four Viking spacecraft that performed so remarkably in the full extended mission will provide material for detailed analysis and study. Much of this study will try to relate what we have learned about Mars to our knowledge of the other planets, particularly the earth; other efforts will try to decide on the next step in exploring Mars.

The Vikings measured physical parameters of Mars more precisely than was possible before. The sidereal rotation rate (rotation period relative to the stars) was established as 350.891986 degrees per day; the spin-axis celestial coordinates of right ascension and declination are 317.340 and 52.710 degrees respectively for the 1950 epoch. Thus the star Deneb in the constellation of Cygnus is the pole star of Mars. It is brighter than earth's pole star. The period of the Martian sidereal day is 24 hours, 37 minutes, 22.663 seconds. The solar day is 24 hours, 41 minutes, 25.36 seconds. Because of Mars' motion around its orbit, a point on the planet's equator takes a little longer to move around and face the sun than it does to move around and face a given part of the star sphere.

Viking also measured the distance from earth to Mars to within an error of 85.3 feet (26 m) compared with 1.24 miles (2 km) before the mission.

The landers established precisely the latitude and longitude of two locations on the planet. That of lander 1 was further refined by observation of an eclipse of the sun by Phobos (figure 9.1). The shadow, about 56 miles (90 km) long, was carefully timed as it swept over the landing site, and the lander took pictures of it. The shadow analysis is expected to fix the location of the lander to within 650 feet (200 m).

In Chryse Planitia, lander 1 touched down at 22.483 degrees north latitude, 47.94 degrees west longitude, at 4:13 P.M., Mars local time (5:12 A.M. PDT) on July 20, 1976, 1.15 miles (1.85 km) below the mean or average level of the Martian surface. In Utopia Planitia, lander 2 touched down at 47.968 degrees north latitude, 225.71 degrees west longitude, at 9:06 A.M., Mars local time (3:58 P.M. PDT) on September 3, 1976, 1.82 miles (2.93 km) below the mean level of the surface.

The highlights of the expedition to the Red Planet started with the landings themselves. Because Mars was so far away that messages took 18 minutes to

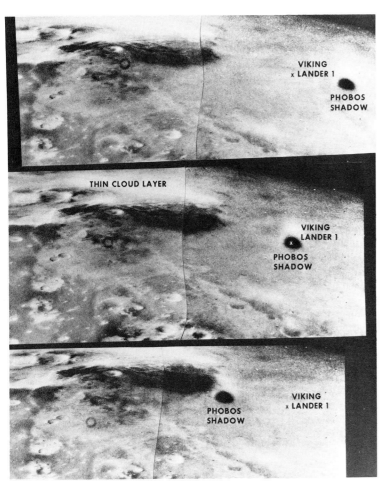

9.1 Viking orbiter 1 took pictures as part of an experiment to locate the position of the lander on Mars using the shadows of the moons Deimos and Phobos. These pictures are the first (*bottom*), middle, and last in a 40-picture sequence of the movement of Phobos' shadow during a 3-minute period. At this same time the lander took pictures on the surface as the shadow moved across it.

Picking landing sites on Mars was rather a terrifying experience for those involved. The surface was much more rugged than expected, and a rock even a foot high could puncture the bottom of the lander. New information and ideas for site selection came in from Mariner pictures, from astronomers, from theoreticians, from infrared orbital data, and from radar observations using big antennas on earth. All these data caused many changes in thought about where it was safe to land. The tremendous adaptability of the Viking expedition was demonstrated many times in finding and going to safe havens on the Martian surface.

Adaptability was, indeed, the trademark of Viking. The mission controllers and the scientists constantly responded to new data coming in from the spacecraft and elsewhere, and constantly updated the expedition. Useful suggestions came from all up and down the organization chart.

Major initial problems turned into minor problems. The hazards of heat sterilization before launch were converted into the advantage of high reliability of the spacecraft. The development of computer software (programs) had been downplayed during early planning, but dedicated effort developed the software needed to interpret the volumes of data from all the instruments of the spacecraft.

One scientist commented that "initially [the mission] was a phase of terror, then a phase of exhaustion, then one of what has happened and what do we do next?"

The Viking spacecraft made many important measurements and discoveries.

Nitrogen, argon, krypton, and xenon were discovered in the atmosphere of Mars, and their proportions measured very accurately. The relative abundances

reach it, mission control could not direct the landings from earth. Each spacecraft had to be released from its orbiter and from control by earth and had to discover for itself, fully automatically, where the surface of Mars was, what the atmosphere was really like and how much it slowed down the fall to the surface, and when to operate the various stages of the landing sequence. It had to turn on its descent engines and land. The Russians had tried unsuccessfully four times to do this. Both the Viking spacecraft touched down safely and in excellent condition.

Epilogue

of the isotopes of carbon, nitrogen, oxygen, and argon were established—important to developing theories of how the atmosphere of the planet originated and subsequently evolved.

The meteorology of Mars at two places on opposite sites of the planet was observed, and the important variables of temperature, pressure, wind velocity, and wind direction were measured for almost a Martian year. Changes in concentrations of atmospheric water vapor on Mars were measured globally on a daily, seasonal, and geographical basis. Discoveries were made about where the vapor concentrates and when. Temperature profiles were obtained for large areas of the surface on a geographical and daily basis, and vertically through the atmosphere over two regions of the planet.

It was established that the permanent polar caps of Mars consist of water ice, not carbon dioxide, and that the polar regions have experienced periods of climatic changes as seen by layering of the underlying terrain. Layering of terrain seen in Mariner pictures of the south pole was shown to occur also in the north pole region.

Viking discovered the largest known dune field in the solar system, adding to the other records of Mars: the largest volcano, the largest canyon.

The abundances of major elements in the Martian soil were measured for the first time. Viking discovered a complex chemistry of the Martian soil, but whether or not it resulted from biological activity could not be determined. Surprisingly, Viking discovered that there were no organic materials in the soil of Mars, even though the detector used was more sensitive than anything previously developed for use on earth.

Viking discovered that the channels, first seen on the Mariner pictures, are common all over the planet.

Viking also discovered mysterious groves on the surface of Phobos that suggested a layered structure, and found that this moon has a relatively low density.

The geological evolution of Mars seems somewhat clearer now than it had after Mariner 9, but opinion is hardly unanimous. The history of Mars appears to encompass four distinct periods. The first period, after accretion of the planet, was the formation of the ancient cratered terrain with some almost immediate fluvial activity. The second was the volcanic destruction of much of this ancient terrain during the formation of extensive lava plains. Thus far, except for fluvial activity, evolution was similar to that on the moon and Mercury. It may have been very similar to evolution of the earth and Venus. The third stage was one of intense activity: the chaotic terrains were formed, possibly by melting of an ice–rock mixture; faults developed and canyons were formed; water flooded over the surface; and the volcanic Tharsis bulge swelled high above the mean level of the Martian surface, spreading great fractures for thousands of miles. The cause of this upwelling is unknown. The final stage, which extends to the present, is a relatively quiet period during which volcanic deposits slowly accumulated in the Tharsis region, and erosional forces continued to mold the surface. During the latter part of this period the polar laminates achieved their present form, and the laminated terrain shows evidence of a number of climatic changes during this stage.

Relative ages have been established for several areas of the planet by counting craters. These ages assume, however, that the rates of cratering followed the patterns of the moon and Mercury, which may not be true because Mars is close to the asteroid belt. The crater counts show that many fluvial channels are extremely old and that lava has flowed from the big volcanoes until comparatively recently.

The variety of volcanic activities that have taken place on Mars proved surprising: great volcanic plains with extensive lava flows, huge volcanic sheields, and great volcanic centers in the uplands which are quite different in shape and form from the shield volcanoes. The oldest volcanoes on Mars probably go back 3.5 billion years, the youngest are so young that no impact craters are visible on them—how young is still not known.

The great Tharsis plateau of Mars is extremely unusual by terrestrial standards. Nor is there anything like it on the moon or Mercury. It is the area of the biggest gravity anomaly (positive) on Mars, and is the highest known plateau in the solar system, almost twice as high as the Tibetian plateau. Thin lava flows cover parts of the plateau, but sticking out at its top are ancient crustal materials that have been lifted some 6 miles (10 km) above the mean surface level of Mars. That this great mass still stands high above the mean level implies a thick rocky crust to support it, because the Tharsis plateau consists of dense rock which causes the gravity anomaly. Another possibility is that this mass is held up by dynamic forces still acting within Mars through a thinner crust, that Mars is by no means a dead planet. Two seismic events recorded within a few months of the seismometer's starting operation in Utopia appear to confirm that Mars is still active today.

The orbital pictures confirm and amplify the importance of water in the history of Mars. Mars has generally been cold throughout its history, with relatively few periods of warm climate. There seem, however, to have been many episodes of water flowing on the Martian surface. Many channels may be the result of local rainfall.

If the results of the biology experiments at Chryse and Utopia can be demonstrated in terrestrial laboratories to be due to chemistry and not biological chemistry, Viking will have proved that life does not exist everywhere on Mars. It will not have proved that there is no life on Mars. At the time of writing, the results of the biological experiments except for some of the pyrolytic-release results can be explained on the basis of a photochemically activated soil held away from chemical equilibrium because of the low temperatures and the lack of water. Add water and warm this soil and chemical reactions immediately occur. But the question of life on Mars is by no means resolved.

The soil chemistry at the two landing sites destroys organic compounds. But some regions of earth's tropics act in the same way. On the edge of the tropical rain forests there are highly sterile lateritic soils that scavenge organics. These soils result from decay of rocks in a rainy climate and they consist of hydrated metal oxides which react with organic substances in the moist climate.

The search for life on Mars must continue if we are to have a definitive answer. But where? And how? Viking-type spacecraft cannot land safely in the most likely places for life on Mars—in the deep canyons, the polar regions, the volcanic calderas, the chaotic terrains. One possibility would be to send an orbiter from which small missiles would be aimed, like darts, to different parts of the Martian surface (figure 9.2). Each would carry a relatively simple set of experiments designed to survive a hard landing, with a spike on the missile's nose to absorb most of the shock. In this way a planetwide survey of several important aspects in the search for life might be conducted at a reasonable cost.

The real search has to take place with more sophisticated spacecraft developed from Viking. One way might be to equip a Viking lander with caterpillar tracks and, when it has landed in a safe area such as Chryse, send it roving over the surface to seek more interesting places. This technique was used in several Soviet unmanned missions to the moon.

Epilogue

9.2 Small missiles launched from a spacecraft in orbit about Mars might survey much larger areas of the surface of the Red Planet in search of conditions suitable for life. Although studies have been made showing the practicality of this type of mission, it is not yet an active program for the future.

Unfortunately, the United States has been gradually going out of the business of planetary exploration. After Viking, only two planetary missions have been planned—a Pioneer to Venus, and a Voyager to Jupiter, Saturn, and possibly Uranus. A probe to Jupiter was rescued from Congressional cancellation by a concerted effort by astronomers and other people who wanted to keep our exploration alive. And astronomers' fortuitous discovery of the rings around Uranus while observing the occultation of a star by that planet shows us that we really are very ignorant about our solar system and the ways planets formed and subsequently evolved. We still have much to learn. Each successful planetary mission has brought important and completely unexpected discoveries that have thrown new lights on our own planet.

There are a few positive indications that the nation is not going to throw away all we have gained in developing a fantastic technical and intellectual capability to explore new worlds. One has been the tremendous public interest in the Viking expedition. Viking did not just look at rocks on Mars; it provided major steps forward in our understanding of climate, meteorology, and geology.

The next opportunity to go to Mars and continue our exploration is in 1984. We have missed this opportunity by not starting a program now, even though such a program would have created many jobs and stimulated the economy as earlier space programs have done. The mission would have been a Mars Rover (figure 9.3). The technology to make this new expedition is available, and the scientific returns would be high. But the cost also is high (about four times the current annual subsidy for tobacco over the

Epilogue

9.3 The future exploration of Mars requires roving vehicles that can search out interesting areas on the Martian surface. This artist's concept is for a Viking type of lander equipped with caterpillar tracks instead of footpads. Such a vehicle could be set down at a relatively safe landing site such as Chryse or Utopia and then be sent to roam over the Martian surface to more interesting areas. The really productive exploration of Mars will have to await a manned expedition, however, and this is unlikely before the beginning of the next century.

period to 1984), so the program was not started. Yet how realistic is it to place a cost restriction on solving the deepest human questions concerning our genesis within the solar system, when we spent more on encouraging the production of known carcinogens? How we came to be, and why we are here on earth, and where we might be going, are fundamental questions that people have sought to answer since the dawn of intelligent life on this planet. Today we know that complete answers are not likely to be found on the earth. We have to look at and on other worlds in a search to find out why they are not like our world and why they followed different paths of evolution.

There is no doubt that we will have the technology to send a manned mission to Mars when the space shuttle is operational and able to transport people and equipment inexpensively into space. The manned expedition would be assembled in earth orbit, possibly as a fleet of spacecraft. On reaching Mars, these spacecraft would be placed in orbit about the planet and send landing craft to the surface (figure 9.4). Such an expedition is not likely to take place until

early in the next century, unless some completely novel discovery is made about the Red Planet by Soviet unmanned missions that are expected in the coming years, building on the knowledge gained by Viking, or by our own Rover mission that might still be sent to Mars during the late 1980s if we decide to build on Viking.

Mars has intrigued mankind over the centuries. It may offer potential for meeting mankind's future needs. During the summer of 1975 a group of scientists from various universities met at NASA's Ames Research Center, in Mountain View, California, to in-

174

Epilogue

9.4 This could be the way man will go to Mars in the late 1990s if we start planning soon for such a mission. Solar-sail interplanetary shuttles each could carry several tons to Martian orbit and be used to land two manned capsules, various rovers and processors to mine and study the Martian surface. Before astronauts arrived, a nuclear power station also would be landed. Continuing research and development might make possible a 3-year mission of this nature to Mars by a 4-man crew within two decades, according to a JPL scientific study group.

vestigate whether it might be possible to use Mars as a habitat for terrestrial life, including man. No insuperable limitation of Mars' ability to support a terrestrial ecology could be identified. The study concluded that an adequate oxygen and ozone-containing atmosphere could be created on Mars through the use of photosynthesizing organisms from earth. Merely planting these terrestrial organisms on Mars would not be enough; they would take tens of thousands of years to make the planet habitable for man. But if planetary engineering were used to modify the climate of Mars by melting the permanent polar caps, and genetic engineering were used to modify some of the terrestrial life forms to make them more efficient oxygen producers tailored to the present Martian environment, the change to Mars could be speeded up considerably.

And as the glaciated terrain warmed up, Mars might flow again with water. But this time, because of the intervention of men and women from earth, the plains of Chryse and Utopia, of Hellas and Arcadia could hold their water in vast new oceans. Mars, no

longer a baleful Red Planet associated with the god of war, could become a smaller version of the blue-green jewel of earth.

And man, made in the image of the Creator, would have created the first of many new worlds for his descendants.

The exciting thing I have learned from Viking is that if we really want to, we can live our dreams: from a magazine article on a Martian probe, to the Viking landing on Mars; from the Viking landing on Mars to a future limited only by our own imaginations.

Epilogue

Index

Index

Index

tectors, 40, 41, 61, 117, 122, 126; management, 38; memory, 62; meteorology, 56; organization, 66; path to Mars, 45, 51; power, 46, 48, 49; redundancy, 51; science, 51, 53; seismology, 56; sterilization, 49
Viking orbiter, 37, 47, 51; battery of orbiter 1, 64; experiments, 52; power, 51; problem with orbiter 2, 88; rocket engine, 51; scan platform, 51

Viking science steering group, 38
Vishniac, Wolf, 39, 41, 61
Volcanic: features, 153; period, 152; plain, 158; wind theory, 25
Volcanoes, Martian, 33, 57, 69, 106, 136, 152, 155, 172

Wabash crater, 76
Walks around Mars, of orbiter, 105, 131
Wallace, Alfred Russel, 5
Water, 4, 8, 12, 15, 16, 28, 31, 32, 35, 53, 56, 74, 76,
83, 122, 128, 134, 135, 139, 144, 145, 150, 166, 172
Wave clouds, 136
Wave of darkening, 7, 28
W cloud, 136
Weather: conditions, 104; reports, 93, 102; stations on Mars, 56
Weight of microorganism, 121
Welles, Orson, 4
Wells, H. G., 4, 5
Wind, 104, 105; erosion, 35, 145; speed, 105

Wohler, Friedrich, 4
Wolf trap, 39
Work shifts, changing, for Viking, 66

Xenon, 98
Xenon arc light, 115
X-ray fluorescence spectrometer, 58, 93, 100

Young, S., 12
Young, Thomas, 70, 83, 87, 91, 95, 126

Zond-2, 29

Index   181